Formal Methods in Systems Engineering

Formal Methods in Systems Engineering

Edited by Peter Ryan and Chris Sennett

Springer-Verlag
London Berlin Heidelberg New York
Paris Tokyo Hong Kong
Barcelona Budapest

Peter Ryan, BSc, PhD
Chris Sennett, MA, DPhil

Defence Research Agency
St Andrew's Road, Malvern
Worcestershire WR14 3PS, UK

ISBN 3-540-19751-6 Springer-Verlag Berlin Heidelberg New York
ISBN 0-387-19751-6 Springer-Verlag New York Berlin Heidelberg

Library of Congress Cataloging-in-Publication Data applied for

Typesetting: Camera ready by editors
Printed by Antony Rowe Ltd., Chippenham, Wiltshire
34/3830-543210 Printed on acid-free paper

Preface

Digital systems are becoming ever more pervasive in our modern, technological society. More and more aspects of our lives rely increasingly heavily on such systems. Many of these applications are of great complexity and criticality. This criticality may be of a safety nature such as "fly-by-wire" avionics and nuclear power station control systems, or of a security nature such as financial transaction systems and databases of sensitive information. In any case it is of great importance that these systems perform correctly and reliably.

Set against this is the fact that digital systems engineering has not come of age. It is in some ways better thought of as being still in an "arts and crafts" stage rather than as a true engineering or scientific discipline. Nevertheless, systems engineering is rising to meet the challenge of mastering the complexity and providing suitably high levels of assurance of correct and reliable behaviour for such critical applications.

Over the last few decades significant strides have been made in the theoretical foundations of the subject but there are still many open questions. Methods and tools such as structured methods and compilers have also developed. These improve the situation but do not solve the underlying problem.

A comparatively new line of attack comes under the name of "formal methods". In some ways this is perhaps a slightly unfortunate and misleading title but is probably here to stay. It is unfortunate in that it suggests the essence of the approach is formality, that is, comprising elaborate, abstruse symbol manipulations. Certainly such symbol-pushing exercises can play a role but they are by no means the totality of the approach, as the ensuing pages will demonstrate.

Formal methods can be roughly defined as the application of mathematical techniques to the specification and development of digital systems. Such techniques include the use of a mathematical notation to increase the precision of specifications, the use of abstraction and modularity to reduce a problem to "mind-sized" chunks, and the use of analytic techniques to establish that a design or specification has certain properties. More precise definitions can be found in the text, notably in Joseph Goguen's Introduction.

The underlining thesis of this book is the belief that formal methods can make an impact on these problems. They should also help to produce better quality products in more reliable timescales and budgets. By using the more advanced proof and refinement techniques, they can be used to produce systems with high assurance of "correct" behaviour (although this will tend to have a trade-off in terms of production timescales and cost). It must however be acknowledged that on the whole formal methods have yet to make a significant impact on "real", industrial-scale systems engineering. It is an important question, with both technical and social ramifications, as to why this is the case.

In the light of this a workshop (FM'89) was organised and held in July 1989 in Halifax, Nova Scotia to try to assess the state of the art in formal methods and make recommendations for future research and development in the field. The workshop was sponsored by the governments of Canada, the UK and the US and drew together some 50 experts in the field from academia, industry and government in the 3 countries.

One of the key conclusions to emerge from this workshop was that in order to facilitate the transfer of the technology and make formal methods effective in industrial applications it would be necessary to understand how to integrate them with existing software engineering practice, such as quality assurance, structured design, testing and so on. There is a common perception of formal methods as all-or-nothing approaches which exclude other techniques. In fact they can and should co-exist in a fruitful symbiosis with existing practice, and are best viewed as another weapon in the engineer's armoury to achieve quality, dependable systems.

This was chosen as the theme of a follow-on workshop held in September 1991 in Drymen, Scotland with a similar, and partly overlapping, group of participants. This book documents the outcome of the workshop and is intended to give a perspective on the state of the art with regard to the practical use of formal methods in industry. Much of the material was presented as invited talks at the workshop and developed subsequently. Also included are reports from three parallel working groups held during the workshop. These present, as far as can be attained, the consensus of the practitioners on the use of formal methods under the headings:

- Modelling
- Quality assurance
- Design methods

The first of these was aimed at understanding the role of mathematical modelling techniques in specification and design. The second was to study how formal methods could interact with and complement existing quality assurance techniques. The last was similarly intended to investigate the integration of formal methods into existing design methods.

To complete the appreciation of the state of the art, a survey of formal methods tools and applications was conducted throughout 1992 and the results of this are given in the appendix.

After the Introduction by Joseph Goguen chapters 2 and 3 deal with some social aspects of formal methods. Margaret Tierney examines the role of formal methods and the way they are commonly perceived. In particular she discusses the way they have tended to be associated with the high-criticality, defence-type applications. In the third chapter, Donald MacKenzie examines the idea of "proof", in particular formal proof, and shows that it is far from being as cut-and-dried as is commonly thought.

Three further chapters address approaches to the problem of developing large systems. The first of these is by Jean-Raymond Abrial. He is the driving force behind the development of the B-technologies. This is a unified and elegant approach, comprising a notation, method and supporting tools. In this chapter he outlines the ideas behind this approach and some of the lessons learnt. The next chapter is by Pamela Zave. This describes research done with Michael Jackson on techniques for the composition of specifications. By using such techniques, a problem may be broken down into manageable pieces which can then be combined in a controlled fashion. The third is by John Wordsworth, and addresses the often rather neglected issue of ways in which formal methods can be used to improve documentation throughout the development life cycle.

The remaining chapters document various experiences gained in the course of the application of certain formal methods to "real" applications. Anthony Hall draws on his experience in the design and development of an air traffic control system for the CAA. In particular he addresses the issue of what constitutes a "good" specification. Finally Victor Basili examines the nature of "quality", emphasising that QA should be process-oriented rather than product-oriented.

The FM workshops were unusual in the way that they deliberately drew together key players from industry, government and academia. In this way the providers, funders and consumers of the technology all had an opportunity to meet and discuss their hopes and concerns. We hope that this book will make a contribution to the awareness of the potential (and limitations) of an important new technology.

February 1993   Peter Ryan
                                                         Chris Sennett

Acknowledgements

We should like to thank the CSE in Canada, CESG in the UK and NSA in the States for sponsoring the workshop on which this material is based.

Thanks to Stephanie Smit of Elsevier for permission to reproduce the contribution from Jean-Raymond Abrial.

Thanks to Mark Hewitt, Colin O'Halloran and Paul Rathbone for help with the L^AT_EX typesetting.

Contents

1 Introduction

Joseph A. Goguen

I am delighted and honoured to provide an introduction for this book, which documents a workshop held in Scotland in September 1991, to assess the state of formal methods technology. Although I was unfortunately unable to attend the meeting, I had attended its predecessor FM'89 in Halifax, Nova Scotia, and my thoughts were very much with FM'91 while it was taking place; its topic seems as timely now as it did then, and the papers included here seem interesting and relevant.

This preface tries to provide some context, and some perspective. It first asks what formal methods are, and what they are good for; it then describes a specific technique called *hyperprogramming*, about which I happen to be enthusiastic just now, and it concludes with some recommendations for furthering research in formal methods. These reflections are based on an informal position paper that was handed out at FM'89, and (by coincidence) the revisions were carried out during a sabbatical visit to the Technical University of Nova Scotia, in Halifax. The bibliography provides some pointers into the literature on hyperprogramming, since this topic is not otherwise represented in this volume.

1.1 What are Formal Methods?

"Formal" means "having to do with form" and does not necessarily entail logic or proofs of correctness. Of course, the word "formal" is also used in many other senses, but I think that this may be the most appropriate sense for "formal" in the phrase "formal methods." For example, a formal development method gives rules that restrict the allowed forms of program development, and perhaps also the allowed forms of some texts that occur during the process. This does not mean that form is trivial — far from it. Indeed, everything we do is done with form. And everything that computers do is formal, in that definite syntactic structures are manipulated according to definite rules. Usually, we don't do things just to follow the form — we have some purpose in mind, and the formal structures that we use, whether PERT charts, programs, parse trees, or differential equations, have a meaning for us.

For many people the prime example of a formal system is first order logic. This system encodes first order model theory with certain formal rules of deduction that are provably sound and complete. However, our experience with theorem provers shows that it can be difficult to work with this system. Moreover, formal systems that try to capture even higher levels of meaning, e.g., languages for expressing requirements, tend to be even harder to work with, and to have even less pleasant properties. (Later I will argue for a natural hierarchy of levels of meaning, from abstract mathematical objects up towards concrete social values.)

In summary, formal methods are syntactic in essence, but semantic in purpose. In computing science, form does not embody content, but rather encodes it.

1.2 Formal Methods and Mathematics

There has been much confusion about the relationship between computing science and mathematics, and particularly about the relationship between computing science and logic. Unlike numbers, computers have a real physical existence, and so do programs. On the other hand, algorithms and models of computation (such as Turing machines, or term rewriting systems) are abstract mathematical entities. What seems problematic is the relationship between the physical entities and the mathematical abstractions.

In my view, this relationship is the entirely familiar one that the ancient Greeks discovered between bodies in the real world and abstractions in axiomatic geometry. Thus, we can prove theorems about (abstract) points, lines, planes and pyramids, but not about the Great Pyramid of Cheops, whose edges and faces are not very regular. Although we can apply suitable theorems to a physical pyramid, we cannot expect the conclusions to be more valid than warranted by its assumptions. For example, we can compute the volume of the Great Pyramid using a simple formula, but it is only fully accurate for a mathematical pyramid. The situation is much the same with programs and computers. We cannot prove the correctness of a real program running on a real computer. But we *can* prove the correctness of an algorithm, and we can expect a program on a computer to behave as we wish to the extent that the program's execution conforms to the algorithm.

It is an error to conflate mathematical models with the concrete realities they are supposed to represent. Hence it is as much an error to claim that computing science has all the "good" properties of mathematics, as it is to claim that it has all the "bad" properties of the real world. We may call those who make the first error the "Dijkstra school" (everything is provable) and those who make the second error the "Fetzer school" (nothing is provable). Perhaps the excessive optimism of the first helps to explain the excessive pessimism of the second. (See [1] for a balanced discussion of this philosophical controversy, and see [10] for my own views and some further historical context.)

In summary, some parts of computing science are pure mathematics (concerned with ideal algorithms and models of computation) and some parts are applied mathematics (concerned with applying mathematical models to real programs and computers). Later I will argue that some parts are neither of these, but instead have to do with the social context of computing.

1.3 What Good are Formal Methods?

Formal methods are often considered useful only for proving that programs satisfy certain mathematical properties, and also, are often considered too expensive. I believe these perceptions reflect a view of formal methods that is too narrow. In fact, formal methods can reduce time to market, provide better documentation, improve communication, facilitate maintenance, and organise work throughout the lifecycle.

The most difficult problems in system development do not arise in relating algorithms to programs, but rather in determining how a program (running on a computer, which I will largely ignore hereafter) can solve some real world problem, such as preventing the theft of funds or information, detecting enemy missiles, making a profit on the stock market, or ensuring the survival of an organism. The trouble in such examples is that the program must perform in an environment that is enormously complex, rapidly changing, and imperfectly understood. Moreover, the requirement for the program may also be complex, changing, and imperfectly understood; in some cases, it is so bound up with social and/or political issues that even trying to state it with greater precision can engender so much debate about larger issues that general agreement on its meaning is impossible.

The ideal of having accurate mathematical models of the real environment and the real requirement is not achievable for many large, complex systems. This means it is inevitable that methods less than purely formal will play an important role in designing and evaluating real programs. In fact, many informal methods are already important in computing science, starting at a relatively low level with communication media such as graphics, natural language documentation, animation, and audio. I believe that this trend will accelerate as computers become increasingly interconnected with each other and with other parts of the real world, through networks, modems, mice, Fax machines, digital audio chips, radar antennae, video cameras, etc. Finally, how can we be sure that we have formalised the right thing? Or formalised it the right way? Clearly, we need to get outside the formal system in order to make such judgements.

We do not have, and perhaps we never will have, fully adequate theories of the meaning of the information that is encoded in such complex forms as natural language. And the prospects are even less encouraging for fully adequate theories of the significance of such information in larger contexts. But we do have formal rules that describe the structure of the data that encodes many kinds of meaning, and we also have many programs that can manipulate such data for particular purposes. For example, we can divide a message into words and sentences, and count the number of each; we can search for keywords, and do various statistical analyses; and we can do spelling correction to a useful extent (but not perfectly without human help).

There are various *levels* of structure that we might seek to formalise. A computer analysis of a message can be fairly certain about words and sentences, somewhat less sure about spelling, and quite unsure about meaning. We have formal methods that are applicable at each level of structure, but in general, the higher the level, the more important *informal* methods of analysis become. The obvious hierarchical structure has to do with whole/part relationships: words are parts of sentences, and sentences are parts of messages. This is a formal hierarchy, and also a hierarchy of forms. But there is an *informal* hierarchy that is perhaps even more important, whose levels correspond to meanings in wider and wider contexts. For example, one relatively high level of meaning might concern the artistic merit of a text in a certain culture.

I claim that the same is true of program correctness. We can formally verify syntax with considerable certainty. We can formally verify semantic assertions about state with reasonable certainty, although the effort needed seems to grow exponentially with program size. We can hope to better understand some of the interactions of a program with components of some larger systems of which it is a part. But current formal methods do not seem especially useful, for example, in determining the effectiveness of a program in achieving a business plan, or in benefiting society as a whole.

In summary, there is a tendency for formal methods that encode higher levels of meaning to require greater computational resources, and to pass into regions that we do not know how to formalise adequately. However, there often exist some rather "syntactic" methods which achieve a useful expressive power with a modest computational cost.

1.4 The Myth of Control

Managers want to control the programming process; they want to be sure that their product will meet its requirements, and will be finished in time and within cost. This is entirely reasonable, but we all know that in practice, managers often do not achieve this kind of control: programs often don't do exactly what they are supposed to do, and often take much longer and cost much more than estimated. The desire for control motivates the use of "process models" for system development; these try to use abstract programs to describe the flow of tasks and information. The "waterfall" process is an extreme case, where higher-level descriptions are supposed to determine lower level descriptions, in a strict hierarchy from user requirements down to code.

One trouble with such models is that they make insufficient allowance for adaptation, error and creativity. Clients usually do not know exactly what they want; for they do not usually know what is possible or impossible, or how much the various possibilities will cost; also, they cannot foresee all interactions with other systems. (Of course, neither can the programmers.) Furthermore, entirely new capabilities may be revealed during the design or construction processes. All this is frustrating to the programmers, as well as to the managers and the customers. (See [8] for an extended discussion of errors in software engineering.)

It is my view that the total programming process should be as flexible as possible, and that workers at each level should participate in a dialogue with the levels above and below. As advocated by the Brooks report [2], rapid prototyping can help to achieve this goal in some cases. However, it is difficult to construct prototypes directly from requirements, and the usual kinds of prototypes do not help much with many higher level evaluations, such as whether the system should be built at all, or how it should be used and managed, or how it will interact with other systems already in place.

It is sometimes claimed that by using formal methods we can avoid all errors in programming. Even if we interpret this claim in the narrow sense of guaranteeing that some formal mathematical property is satisfied by some formal mathematical specification (as opposed to an informal social requirement being satisfied by a real system operating in a real context), it may still fail to hold in practice, because we can make mistakes; for example, we can neglect to follow the method at some apparently trivial point, with unexpectedly serious consequences; or we can make a syntax error during proof, with the result that we prove a property of the wrong function, or prove a different property than expected, or fail to prove anything at all.

Even worse, attempting to enforce the rigid use of a formal method can be very damaging, by preventing flexibility and inhibiting creativity. Rigid top-down design methodologies do not work for large programs, and can be unpleasant and stifling to use even on small programs. In co-teaching a course at Oxford, I found that some students who believed my co-teacher's assertions that they should be able to get their programs right the first time by using weakest preconditions backwards from a post-condition, lost the confidence that they could ever learn to program at all. This is a great pity.

Under certain conditions, a formal proof of correctness could be worse than useless, by encouraging misplaced confidence that the program will meet its intuitive requirement in its actual operating environment. For example, if a company promotes a formally verified heart pace-maker as infallible, physicians might neglect to provide adequate safeguards.

It is well-known that many of the most important scientific and technological discoveries were accidental (e.g., penicillin), or arose through trial and error (e.g., the light bulb), and I think that we should allow for similar processes of exploration in programming. Formal methods can help with this by providing techniques to ensure the inter-consistency of the many different texts that arise in producing large and complex systems. Note that an exploratory approach can produce even more texts than a more controlled approach, and that these texts may have subtle relationships with prior texts. One approach to managing this complexity is discussed in the next section.

1.5 Hyperprogramming

Large programs have many parts whose interactions and interconnections are under constant evolution during their development. This means that the many texts associated with the program, including its requirements, specifications, code, documentation, accounting information, test suites, and version and configuration files, will be changing constantly. It is highly desirable to provide support for maintaining the mutual inter-consistency of such texts as they change. It is not practical to do this for the *contents* of these texts, but it seems promising to apply formal methods at various levels to their *forms*. For example, consider the manual for an operating system that is to run on several machines, and is frequently corrected, augmented and ported. We would like to ensure that the organisation of the manual remains consistent, and that if part of the program is changed, then the corresponding manual pages are re-examined to see if they also must be changed. This seems fairly straightforward.

But perhaps we can go further. If we associate documentation to the parts of a program, then we can assemble the manual from its parts in the same way that we assemble the program from its components. Furthermore, if the documentation and the program are parameterised in the same way, then we might be able to evaluate a single interconnection statement that would accomplish these two different purposes. In fact, we have already developed a theory of *module expressions* which can serve such a purpose [6]. Their use in programming is called *parameterised programming*, and it can be considered a substantial generalisation of the programming-in-the-large style embodied in the UNIX make statement. One dimension of the generalisation is to provide powerful facilities for both generic modules and module inheritance; the former also allows us to specify semantic properties of module interfaces.

Parameterised programming has been implemented and extensively used in the functional programming language OBJ [14, 3], and has also been suggested for Ada and other languages [4]. *Hyperprogramming* is the extension of this approach to texts other than programs [7]. For example, it could be used to combine graphical illustrations with written texts, to assemble a spoken explanation from parts and then "execute" it with a speech chip, to produce program animations from specifications of program parts, and perhaps even to combine such animations with speech to form "movies" that illustrate program operation.

Although these considerations motivated the name "hyperprogramming," its most important application might be the coherent integration of the many different components of a large software development project. Current practice does not support the integration of rapid prototyping with the evolution of specifications and code, nor does it support consistency checks between such texts as specifications, test suites and code. Moreover, accounting and management information are usually handled quite separately from code, and documentation is only developed after coding is completed. Hyperprogramming could integrate all these diverse aspects in a uniform way that guarantees certain important kinds of consistency.

1.6 Recommendations

The suggestions in this section are based on my own experience. Of course, this means that they are biased. But I think this should be considered a strength rather than a weakness, as long as the source is clearly indicated. Except for the first list, these recommendations can also serve as a summary of the preceding discussions. I begin with some observations that have to do with general funding policies:

- Educational expectations are too low in computing science, and in particular, they are lower than in other engineering disciplines. Resources should be applied to raising both the mathematical preparation (especially in logic and algebra) and the practical experience of computing science graduates. Formal methods should be taught in the universities. So should informal methods.

- Open dissemination of basic research results and of experimental systems is essential in order to obtain the best use of research funding by maximising the discussion of critical issues. Unfortunately, some very good work has not received the attention that it deserves, because of restrictions imposed by its sponsors.

- It is difficult to get funding for innovative ideas that require the development of prototypes, because the funding programs with sufficient money tend to have goals that are excessively narrow and short-term.

- Funding is often unstable and subject to excessive delays. As a result, it is difficult to hire and keep good people.

Next, I list some fairly specific research topics on the border between formal and informal methods. As previously argued, this area seems very important for software methodology. Although short term practical results seem unlikely, I believe that important basic results can be obtained by people who are proficient in both formal and informal methods. The informal (or perhaps one should say "semi-formal") methods that seem most relevant come from the social sciences, and include discourse analysis, socio-linguistics, and ethnomethodology.

- It seems likely that the dialogical processes between clients and designers that result in requirements could be formalised to a certain extent, and that this could, at the very least, result in more realistic expectations about what can be accomplished in this stage of the development of complex systems; see [9] for some related discussions.

- A linguistic study of the relationship between requirements texts and the texts produced at lower levels, such as designs and specifications, might yield formal structures that would facilitate these important transitions in the program development process; see [11]. In particular, it would be interesting to know where misunderstandings most often arise in current processes.

- It would be interesting to study the integration of various kinds of text in various media, to see what constraints must be satisfied to ensure that the intended relationships are actually perceived by users.

- I believe that the organisational methodology of formal methods can yield benefits without requiring full mathematical rigour. For example, one might keep track of proof obligations without doing all proofs. In particular, one might use multiple levels of formality; for example, levels might include full mechanical proof, informal mathematics proof, back-of-envelope proof, intuition and hope. Then one could do a critical path analysis of the arguments that support a given requirement, and the formality of items on the critical path could be increased if desired.

- It would be important to study how to integrate formal methods with the wide variety of methodologies and work practices currently used in industry.

Finally, I list some (relatively) specific research topics that lie entirely within the area of formal methods and that I think could yield substantial advances within a medium-term time frame:

- The integration of specification, prototyping and theorem proving. This could be done by using an executable specification language that is rigorously based upon logic; indeed, every execution of a program in such a language is a proof of something, and if the language is rich enough, it could be a proof of something interesting. We have done some experiments in hardware verification using the OBJ3 [14] and 2OBJ [13] systems which suggest that this is a promising research direction [5, 15].

- The development of semantically rationalised object-oriented programming languages. For example, an integrated functional and object-oriented language (such as FOOPS [12]) could be useful for many purposes, including structured verification, and integration with the so-called structured design methodologies, like those of Jackson and Yourdon.

- Developing support for an entire object-oriented lifecycle; this is related to the development of rationalised semantics for object-oriented programming, but would also be useful in connection with existing languages like Ada. This would involve developing CASE tools for object-oriented verification, and for object-oriented requirements.

- The development of prototype hyperprogramming systems. This might include integrating the graphical conventions used in structured development methods. It would also be interesting to study the automatic generation of graphical representations of abstract data types from their algebraic specifications; it seems likely that this could be extended to the automatic generation of animations.

Of course, the pursuit of these research goals will not solve the philosophical problems discussed earlier in this preface. However, the philosophical considerations do tend to support the belief that such research goals may be both important and feasible.

I hope that no one will misunderstand my interest in informal methods, my criticisms of current methodologies, or my general philosophical positions, as being critical of formal methods. On the contrary, I believe that we may be on the threshhold of a golden age of formal methods, provided that we neither interpret the field too narrowly, nor expect it to achieve the impossible. Also, there seem to be many areas where formal methods can be synthesised with informal methods to solve important problems in the development of large systems.

References

[1] Jon Barwise. Mathematical proofs of computer system correctness. Technical Report CSLI-89-136, Center for the Study of Language and Information, Stanford University, August 1989.

[2] Frederick Brooks *et al.* Report of the Defense Science Board Task Force on Military Software. Technical Report AD-A188 561, Office of the Under Secretary of Defence for Acquisition, Department of Defence, Washington DC 10301, September 1987.

[3] Kokichi Futatsugi, Joseph Goguen, Jean-Pierre Jouannaud, and José Meseguer. Principles of OBJ2. In Brian Reid, editor, *Proceedings, Twelfth ACM Symposium on Principles of Programming Languages*, pages 52–66. Association for Computing Machinery, 1985.

[4] Joseph Goguen. Reusing and interconnecting software components. *Computer*, 19(2):16–28, February 1986. Reprinted in *Tutorial: Software Reusability*, Peter Freeman, editor, IEEE Computer Society, 1987, pages 251–263, and in *Domain Analysis and Software Systems Modelling*, Ruben Prieto-Diaz and Guillermo Arango, editors, IEEE Computer Society, 1991, pages 125–137.

[5] Joseph Goguen. OBJ as a theorem prover, with application to hardware verification. In V.P. Subramanyan and Graham Birtwhistle, editors, *Current Trends in Hardware Verification and Automated Theorem Proving*, pages 218–267. Springer, 1989. Also Technical Report SRI-CSL-88-4R2, SRI International, Computer Science Lab, August 1988.

[6] Joseph Goguen. Principles of parameterized programming. In Ted Biggerstaff and Alan Perlis, editors, *Software Reusability, Volume I: Concepts and Models*, pages 159–225. Addison Wesley, 1989.

[7] Joseph Goguen. Hyperprogramming: A formal approach to software environments. In *Proceedings, Symposium on Formal Approaches to Software Environment Technology*. Joint System Development Corporation, Tokyo, Japan, January 1990.

[8] Joseph Goguen. The denial of error. In Christiane Floyd, Heinz Züllighoven, Reinhard Budde, and Reinhard Keil-Slawik, editors, *Software Development and Reality Construction*, pages 193–202. Springer, 1992. Also in *Four Pieces on Error, Truth and Reality*, Programmming Research Group Technical Monograph PRG–89, October 1990, Oxford.

[9] Joseph Goguen. The dry and the wet. In Eckhard Falkenberg, Colette Rolland, and El-Sayed Nasr-El-Dein El-Sayed, editors, *Information Systems Concepts*, pages 1–17. Elsevier North-Holland, 1992. Proceedings, IFIP Working Group 8.1 Conference (Alexandria, Egypt); also, Programming research Group, Technical Monograph PRG–100, March 1992, Oxford.

[10] Joseph Goguen. Truth and meaning beyond formalism. In Christiane Floyd, Heinz Züllighoven, Reinhard Budde, and Reinhard Keil-Slawik, editors, *Software Development and Reality Construction*, pages 353–362. Springer, 1992. Also in *Four Pieces on Error, Truth and Reality*, Programming Research Group Technical Monograph PRG–89, October 1990, Oxford.

[11] Joseph Goguen and Charlotte Linde. Techniques for requirements elicitation. Technical report, Centre for Requirements and Foundations, Oxford University Computing Lab, 1992. To appear, *Proceedings, Requirements Engineering '93*.

[12] Joseph Goguen and José Meseguer. Unifying functional, object-oriented and relational programming, with logical semantics. In Bruce Shriver and Peter Wegner, editors, *Research Directions in Object-Oriented Programming*, pages 417–477. MIT, 1987. Preliminary version in *SIGPLAN Notices*, Volume 21, Number 10, pages 153–162, October 1986.

[13] Joseph Goguen, Andrew Stevens, Keith Hobley, and Hendrik Hilberdink. 2OBJ, a metalogical framework based on equational logic. *Philosophical Transactions of the Royal Society, Series A*, 339:69–86, 1992. Also in *Mechanised Reasoning and Hardware Design*, edited by C.A.R. Hoare and M.J.C. Gordon, Prentice-Hall, 1992, pages 69–86.

[14] Joseph Goguen and Timothy Winkler. Introducing OBJ3. Technical Report SRI-CSL-88-9, SRI International, Computer Science Lab, August 1988. Revised version to appear with additional authors José Meseguer, Kokichi Futatsugi and Jean-Pierre Jouannaud, in *Applications of Algebraic Specification using OBJ*, edited by Joseph Goguen, Cambridge, 1992 (?).

[15] Victoria Stavridou, Joseph Goguen, Andrew Stevens, Steven Eker, Serge Aloneftis, and Keith Hobley. FUNNEL and 2OBJ: towards an integrated hardware design environment. In *Theorem Provers in Circuit Design*, volume IFIP Transactions, A-10, pages 197–223. North-Holland, 1992.

2   Formal Methods of Software Development : Painted into the Corner of High-Integrity Computing?

Margaret Tierney

There is an old joke about a party of tourists, lost in the back roads of the West of Ireland, who ask a farmer for directions to Castlebar: "Well", says he, "If I were you, I wouldn't start from here at all." The same might be said to formal methodists attempting to gain a general legitimacy amongst mainstream software developers. The emerging techniques, tools and management practices used for the formal specification and verification of software signal an approach to design, development and implementation which is frequently at odds with existing practices. This paper briefly reviews the reasons why software developers and users, by and large, resist formal methods.

That said, there is one niche in software engineering where formal methods already enjoy considerable success. I refer, of course, to the field of high-integrity computing, where user requirements for safety, security or reliability are sufficiently critical to dominate the time and cost constraints which characterise every development project. Where the quality of the software is the main event, formal methods have justly garnered serious attention over the last decade. The hypothesis raised here is whether this very success in the niche of high-integrity computing may compound, rather than assuage, reluctance to embrace formal methods more generally.

Donald MacKenzie, Eloina Pelez and myself are currently investigating the generation and use of formal methods, under the ESRC-funded Programme on Information and Communication Technologies (PICT). For us, what is interesting about this nascent technology is how it comes to be shaped by the social relations of its production; how it comes to bear the marks of those who fund it, build it, buy it and exploit it. Such a research agenda depends on exploring detailed case-studies of formal methods R&D and implementation (many of which lie in safety- and security-critical fields). We are still at the early point in building up a composite picture of formal methods, and our fieldwork is only now in process. Thus, the hypothesis raised here is tentative and partial.

For one thing, these notes do not distinguish systematically between different kinds of formal method. There is a world a difference between, say, formal specification and automated theorem-proving, though both fall under the general rubric of formal methods. In addition, formal methods are applied to different components of a system which introduces important variations in purpose and use. Again, these are not explored here. Nonetheless, there may be some value in treating formal methods for software development as a single entity.

Not least, this single entity perception of formal methods is, most likely, how outsiders view the field. Those furthest from a field of endeavour will not be aware of the specific potentials and uncertainties which the field's insiders - the core set - are privy to [2, 8]. There are two lines of inquiry which spring from this:

1. Does the dominant rationale for formal methods emphasise the difference of formal methods from current practice? If common ground is

apparently limited, to what extent is resistance to formal methods expressed as a "formula reaction", because most developers cannot see the detail of what formal methodists do?

2. Is that detail especially difficult to display to outsiders? Does the field of high-integrity computing, as the dominant context for the use and practice of formal methods, itself contribute to the reproduction of resistance? Do the current practices of formal methodists - their work organisation; their information networks; their highly specialised skills and occupational language - work to perpetuate the situation?

2.1 The Dominant Rationale for Formal Methods

Within software engineering, formal methods harness the model-building and reasoning capacities of mathematics to offer a particularly promising way of gaining greater intellectual control over the software design process. Mathematical notations are used to specify or model parts of the system at one or more stages of its development. Mathematical reasoning may then also be brought to bear to show the correspondence between two or more models. In common with any innovation where extensive empirical evidence of the innovation's value is hard to come by, it is an idea of formal methods which first needs to be sold. Champions must assert the value of these techniques to developers, funders and clients alike in a manner which makes the transition to formal techniques become demonstrably worthwhile [12].

In search of such a rationale, formal methodists have tended to arrange their arguments around the over-riding necessity to understand what it is we are building, when we build software. Improved intellectual visibility of design assumptions and development processes is what is promised, and often delivered, through using formal methods. The corollary of this rationale is that productivity savings will, eventually, emerge as a natural outcome. "Control through understanding" is billed as the most crucial criterion for building cost-effective software which is fit for its purpose. Everything else is seen to follow from this: capturing the essential properties of a specification, regulating the clarity and transferability of designers' work, achieving cost savings over the entire life cycle, and so on.

However, it is precisely around everything else that conventional software developers take issue with formal methods. Much of the resistance is articulated around the notion that the formal approach demands a break with all that has gone before, without offering a means of accommodating to the pragmatics of "real world" software product and labour markets [1, 6]. The real world, in this context, refers overwhelmingly to addressing developers' ability to deliver the goods, in the face of increasing user demands, using whatever skills, techniques and tools they currently possess [11]. It is not that software quality is not an issue, for it is. Rather, in the real world, productivity cannot be made subservient to quality, for profit is premised on speedy delivery of adequate software at lowest possible cost. The reasons given for reluctance to endorse formal techniques are most usually expressed as aspects of this productivity problem.

2.2   Some Pragmatic Objections to Formal Methods

Formal methods pose difficulties in relation to the existing skill base of software-related labour. The majority of designers and practically all clients are not mathematically trained or, even if so, have the traditional engineer's background in the calculus rather than in the areas of logic which are of greater relevance to formal methods. To conceive of software design in terms of its algebraic and logical properties requires some mathematical sophistication on the part both of the designer herself and the user who is involved in specifying that design [7]. This shortage in the existing software skill pool cannot be quickly turned about, for it implies a long-term infrastructural re-orientation of software engineering national training programmes. There are three sorts of critical reactions which stem from this:

1. Many software engineers see any move to incorporate the application of mathematics to systems development as highly desirable in undergraduate education. Indeed, formal methods are particularly promising since, almost uniquely in software engineering, they offer a means of uniting an underlying theory of software with an emerging collection of usable techniques. However, given the current absence of such an infrastructure, the cost of releasing staff for adequate training is high. Not only that. Once trained on, say, a particular notation, the slow climb through the learning curve bites a second time into immediate productivity.

2. Many remain to be convinced that the underlying mathematics of software design, which formal methods make explicit, are the appropriate medium for eliciting and capturing client's requirements. Current systems development is user-driven, in that "user satisfaction" has become the major criterion in evaluating the success of a project [4]. In this respect, as Pelez notes, it is "a colossal must to hide the machine's mathematical nature as much as possible" [10] p.223/4. Mathematical notations are "positively manager-bellicose" [9] p.46. Thus, a developer using formal methods must take additional pains to translate specifications into a form which is comprehensible to clients, and the use of these techniques is likely to generate new specialisms, and new costs, in specifying and implementing a formally-modelled piece of work [9].

3. Given the historical concentration of formal methods R&D on proving program correctness, the uncompromising mathematics of these techniques are seen to undermine existing software engineering expertise, that is of being able to choose pragmatically from an array of techniques borrowed from other engineering disciplines. Engineering know-how, dependent on tacit judgement gained through experience, honours efficient solutions rather than correct ones. To the extent that formal methods are widely understood as demanding the opposite choice – as witnessed, for example, by industry's dismay about the "banned practices" in the initial draft of the Ministry of Defence's 00-55 standard [14] – formal methods are seen as subverting, rather than strengthening, existing software engineering skills.

A development process centred on formal methods pushes its costs upstream: the specification and design stages cost more, while implementation and maintenance costs are presumably reduced. Indeed, this very change of emphasis from down-stream to up-stream is one of the major promised benefits of formal methods, for they are better equipped to catch ambiguities of specification which become increasingly costly to correct the further down the life-cycle the software goes. However, the up-stream focus of formal methods is understood as demanding radical and expensive change to the current organisation of the software development process

4. Formal methods presuppose an infrastructural revolution in how cost-benefit analysis (CBA) is done in budgeting for new projects. Existing CBA practices are biased against formal methods, because of the high salience they give to training and development costs, and the low salience to training benefits and maintenance costs. In addition, the project managers responsible for commissioning or building software are not normally the same people - and sometimes not even the same departments - responsible for maintaining the software later. On this account, they have a problem justifying the higher initial cost of projects premised on the formal approach, when the benefits are difficult to assess in advance, and anyway, probably fall into somebody else's patch (Quintas:1992).

5. Extensive documentation of the evolution of a design is an essential prerequisite for the successful use of formal (or, for that matter, informal) methods whose rationale is grounded in changing the fundamentals of software engineering. Formal methods, whose raw material is ideas – models, axioms, and so on – just cannot proceed without documentation. However in many installations, elaborate documentation, despite its obvious virtues, runs against normal practice. With the over-riding time and cost constraints of "routine" software development, where clients demand a product rather than evidence of its process, documentation is an overhead. It seldom makes a direct contribution to further sales or new contracts, yet it eats into developers' time. Thus, as documentation pushes up development costs and lengthens the time taken to complete a piece of work, it is this activity which is glossed in the rush to deliver the goods. Formal methods, amongst other software engineering techniques, challenge this status quo, but do so in the face of the considerable inertia jointly generated by the lack of user demand and developers' need to complete the job.

6. Finally, formal methods pose difficulties on the grounds that they are new techniques, which can only be used effectively in special applications which, by virtue of their one-off nature or their generous budgets, offer some protection against the rough rules of commercial product markets. In a mature market, black-boxed technical artefacts in the form of semi-generic packages and tools offer the developer and client established short-cuts which reduce the time or cost spent on development. In a variation of the aphorism "Nobody is fired for choosing IBM", mature methodologies and tools have an in-built advantage: they are known

about and they have, through experience of use and subsequent redesign, become sufficiently robust to become safe choices. In this respect, formal methods bewilder: there are as many specification notations, languages and tool sets as there are companies selling them. Most are underdeveloped as purchaseable commodities, for each was designed for a limited purpose and, collectively, they do not mesh well with each other [3].

2.3 Dissolving Resistance to Formal Methods?

In respect to each of these commonly-cited reasons for resisting formal techniques, formal practitioners, as insiders, recognise what is over-generalised about the way resistance is articulated. These "formula" responses may misrepresent formal methods or may overstate the degree to which they demand revolutionary change if they are to work. For instance, [5] argues that they are "myths", to each of which he provides counter-arguments. Based on evidence of use, he cites how different varieties of formality can accommodate incrementally to existing skills, to an array of applications, and to distinct development phases.

Insiders to the field recognise that formal methods are not homogeneous. Different varieties of formalisms address distinctly different problem domains (from well-made specifications through to rigorous code verification) by means of methodologies and tools, each geared to their particular purpose. Thus, not all formal methods imply the same degree of change to existing skill or organisational structures. In addition, those who use formal methods recognise that they are less "anti-productive" techniques, than techniques which attempt to re-define what genuinely productive software should look like. Thus, even if they are somewhat incompatible with conventional practice, this incompatibility stems from the inadequacies of the conventional short-term view. Quick-and-dirty hacking may produce working software, but the mere fact of it working hardly guarantees its cost-effectiveness.

To dissolve resistance to formal methods, it would appear to be a simple matter for insiders to simply tell what they know; to change rhetorical tack. Indeed, it is common for champions to rebut the idea that formal methods must be an all-or-nothing choice (ie. that unless the outcome is programs which have been proved to correspond to specification, the techniques are not worth the trouble or cost), or a contribution solely geared towards the production of assured software (ie. that unless this is a project's objective, the techniques can happily be ignored). However, the body of evidence for the successful "ordinary" use of formal methods is still relatively slender – particularly with respect to its visible contribution to increasing productivity and/or reducing development costs. Yet without that kind of evidence, it is difficult for insiders to tell what they know, and for outsiders to believe what they hear. If formal methods suffer from the handicap of "myths" [5], it becomes important to understand why those myths are tenacious.

To address this, we need to examine the context which has surrounded the emergence of formal methods as usable techniques; the material conditions from which their imagery springs. Consider, for example, why formal methods have attracted the tag of being all about the pursuit of assured, sound, correct software irrespective of other development considerations – a myth the formal methods community is at some pains to dismantle (eg. [3]). To those outside the community, this has substance. The emergence of the formal approach has been so closely tied with high-integrity application fields that its very history is confirmation of the tag. Conversely, to those within the community, it is their insider knowledge of the uncertainties and limitations of achieving assured software, gained precisely from sustained efforts to address high-integrity problems over two decades, which reveals total assurance as a myth. Either way, the "truth" both of the tag and its refutation is embedded in something else: in this case, the context provided by high-integrity computing.

The current shape of formal techniques for software development embody many of the characteristics of high-integrity computing, where safety or security are the driving demands. In addition, high-integrity computing has shaped the skills and work practices of the formal methods community, where the relations of production between clients and developers encourage close-knit and highly-specialist information-trading networks. The next section of the paper suggests that these characteristics may substantiate the idea that formal methods are not a valid currency for mainstream practice. That is, the dynamics of high-integrity computing make it especially difficult for the "core set" to demonstrate the value of formal methods to those who do not yet use them.

2.4 A Brief Sketch of Formal Methods in High-Integrity Computing

High-integrity projects – ones which, perforce, must prioritise the goals of safety, security or reliability in system design – cover a variety of sectors and application fields, across different national contexts. Indeed, since trustworthiness is a system-based concept, the goal of high-integrity refers to architectures, hardware and software as a whole entity. Thus, not only does the context of high-integrity computing vary enormously across sectors and applications fields, but decisions as to where to locate the bulk of a system's reliability requirements (eg. in its software, or elsewhere) also vary. The brief notes offered here do not distinguish between different kinds of high-integrity projects. The aim is simply to sketch the broad socio-economic constituency of formal methods in high-integrity computing: its projects, its tools, its funders and clients, its developers, its information-trading networks.

2.5 The Projects

Any high-integrity computing project encounters two formidable difficulties. First, the field is new. Though safety engineering is well-established – ancient, even – the organisational and sectoral pressures which make it attractive to locate control over crucial elements of a system within a computer are modern phenomena. On this account, a high-integrity project is one which treads new ground. Its actors must write the rules and build the tools as they go along. Second, despite the extreme uncertainties in doing this work adequately, the price of failure is high. A high-integrity project tackles issues of national or corporate security, or the protection of life. Everything from the conceptualisation of the specification model through to the proving of the underlying algorithms must be understood, built, tested, reviewed, rebuilt and found fit. Not surprisingly, the net effect of these two characteristics is that most high-integrity computing projects are path-breaking in their conception and execution. Thus, the formal methods used within them are embedded in projects which carry all the additional costs associated with uniquely difficult developments.

Though we can assume that, given time, some of the costs associated with projects of this kind will be reduced – formal models and tools (or bits of them) are already being reused – that is not likely to happen quickly or uniformly. The clients who fund formal techniques and tools for use within high-intregrity projects do so in terms of meeting their own immediate specialised needs. That is, unless the formal methods work is being applied to something, it will not be funded. One effect of this is that each project using formal techniques yields a model or language or tool specially geared to its particular purpose. Any additional work on increasing its potential flexibility or general robustness is something which happens as a secondary and slower activity. Since anything except partial re-use (particularly across, rather than within, contracting companies) is frequently an option which is fraught with potentially serious design flaws, it often – ironically – becomes more cost-effective to engage in new development rather than to tailor others' work. In addition, many of the formal techniques and tools developed for clients in defence departments are subject to restrictions on circulation, use or sale elsewhere, especially across national boundaries. Thus, the path-breaking quality of many high-integrity projects is reproduced, rather than assuaged.

Given the typical sorts of requirements demanded by high-integrity funders and clients, formal methods, in this niche, are a set of techniques which are useful for restricting systems: they aid the prevention of the system doing something undesirable. While the clients certainly demand functionality from their systems, their interest in sponsoring formal methods R&D does not generally spring from this aspect of their requirements. Thus, some of the most noteworthy technical and procedural achievements of formal methods (eg. automated theorem-provers; system-centred strategies for dove-tailing formal analysis with testing; organisational routines for independent review of the designer's models and axioms) reflect the preventive focus demanded by these particular clients.

Shaped by this history of use, formal approaches - with their attendant tools and work practices - most visibly prioritise assurance over functionality. They "stand for" the polar opposite of enabling techniques. To those who use formal methods, this may be seen as a false dichotomy. Our research group frequently encounters persuasive arguments that what a formal approach offers is clarity. This clarity of conception, most especially through formal modelling of a specification, can, in principle, be as easily applied to capturing functional requirements as to restricting functionality for purposes of safety or security. However, to those outside the core set, this dichotomy has more substance. The bulk of evidence points to formal methods as techniques whose primary value is restricting software from working wrongly, rather than enabling software to work adequately.

The cost-effectiveness of the formal methods component of high-integrity projects is adjudicated by the client on the basis of its contribution to meeting stringent design requirements. If it manages that, it has paid its way. Thus, the design of formally-based tools is not usually premised on criteria such as the tool's ease of use by (non-expert) others, or by the time or space it needs to complete its tasks. For them to work efficiently, the current state of many verification tools demands that the manipulators of the tools are themselves experts and/or that the amount of memory space available for the tool to work in, is not a critical factor.

2.6 The Formal Methods Community in High-Integrity Computing

2.6.1 The Clients

Without the funding which has been forthcoming from large government departments (most notably, defence) over the last 20 years, it is most unlikely that research into formal methodologies, languages and tools could ever have evolved beyond a few isolated projects; "academic" not merely in location, but also in scale and exploitability. In the case of formal methods, agencies attached to the military (eg. RSRE, CESG, CSE, NSA and its offshoots) have been crucial to their evolution as exploitable techniques. These, together with central government agencies for other sectors (eg. NASA, CAA), constitute the vast majority of the clientele for projects which are premised on the use of formal methods.

As customers, they are untypical. It is they who sponsor R&D in formal methods. Their requirements focus on problems of security or safety: cost-effectiveness (though always desirable) is not the main goal. The trustworthiness of a computer-based system is an attribute which is everywhere inscribed in the process by which that system is built and checked. Thus, what these clients contract for is detailed documentary evidence of that process, as much as for any final working product. For these clients, it is not sufficient for the product to behave, it must be known to behave. The relationship between themselves and their contractors is more likely to be collaborative than wholly transaction-based; developmental, rather than black-boxed, artefacts are acceptable as output. Finally, having a role as regulators, as much as procurers of high-integrity systems, these clients can initiate certification standards which solidify particular "visions" of formal methods within particular fields (eg. the Orange Book with its approved security models and tools; 00-55 with its detailed formalisation of the management of safety).

2.6.2 The Developers

Again, the profiles of those engaged in the design and implementation of formal methodologies and tools depart significantly from the mainstream of software developers. Typically, they are employed in "laboratory" companies which enjoy close associations with universities, and whose purpose is to conduct innovative R&D which may (or may not) result in commercial products. In this sense, the developers are "applied academics", whose design skills are firmly grounded in the theoretical backgrounds from which their interest in formal methods has evolved (eg. mathematics or computer science).

As with academics, many hold doctorates and their career routes have been structured along highly specialist pathways. Their expertise is one of mastery of formidable bodies of theory and its application, acquired over a long period of formal education. As with any specialism, these developers are practised in their own occupational jargon: the terms and symbols and concepts, so troubling and arcane to outsiders, are the necessary disciplinary short-hand through which formal developers progress their ideas with each other.

Like academics also, the organisation and monitoring of formal developers' work is, most typically, peer-oriented. The audience and judge of their work is less their immediate management than their occupational peers (who may be their clients, collaborators or competitors) to whom they routinely circulate a variety of written and verbal reports on their progress. This strategy of independent review of work which is so characteristic of formal methods work in high-integrity computing – yet is so alien to mainstream development practice – trades directly upon an academic rather than an industrial model of work.

2.7   The Information Networks

Given the general characteristics of the key actors, outlined above, and the sorts of specialised design problems upon which they work, it is not surprising that the work-related networks of formal methodists are unusually tightly bounded. It is an occupational community which speaks mainly to itself or to other non-formalist engineers engaged in safety- or security- critical work. They attend the same conferences and workshops; subscribe and submit papers to the same journals; peer review each other's work; visit each other's work sites. While some pockets of this community - notably, those engaged in classified security-critical defence projects - do not discuss their progress in using formal techniques, the whole field of formal methods R&D gains substantially from the tightness of the network.

What can be efficiently traded through a network of peers, who know each other or of each other and who hold certain basic problems in common, is detail. Indeed, as noted earlier, it is knowing the detail which enables the core set – the insiders – to recognise both the potentials and the uncertainties of designing and using formal methods. The flip side, of course, is that a tight network is exclusive: learning from the detail assumes that the members of the network engage in the same sorts of problems, speak the same language, share the same basic assumptions. This pattern of information flow is, again, much more akin to academia than to industry. User managers, project managers and programming staff engaged in mainstream software development cannot easily participate, even if they wished to.

2.8   Does the Use of Formal Methods within High-Integrity Computing Perpetuate its "Myths"?

This paper opened by summarising some of the common themes as to why formal methods cannot be readily incorporated into general software engineering practice. If the techniques cannot be seen to address issues of productivity and cost-effectiveness, except at the price of massive infrastructural upheaval, these themes are likely to be re-played. Indeed, evidence of the use of formal methods in the 'exceptional' case of high-integrity computing is simply read as evidence that, in this niche, these normal constraints can be waived [13, 9]. The question, then, is how the specialness of this computing niche may itself subvert the attempts of the core set to show how formal methods can earn their keep in any application field to which they are applied.

In the sketches of the last section, we saw that the use of formal techniques in high-integrity computing is frequently embedded in projects which, for a host of reasons regardless of their formal methods component, are expensive anyway. Thus, the image of formal methods as being inherently expensive techniques which require advanced specialist skills is compounded by evidence of their use in high-integrity applications. The obvious policy issue is to ask how formal methodists, who seek to colonise new application fields, can extract out some credible estimates of what formal methods "really" cost, or what kind or level of skill they "really" demand.

In this respect, the typical shape of high-integrity projects provides little help. Most of these projects cannot readily be compared to anything else. Possessing unique characteristics, where the formal component is only one in a wider array of design and testing strategies, it is hard to show that the project would have cost more, or would have taken longer to finish, if the formal approach had not been used. While, say, an insider may know from experience that the formal specification of a problem domain will routinely uncover potential errors early on in the life cycle, it is difficult to extract out this argument so that it is plausible to anyone outside the immediate design team, or the larger network of peers.

In addition, the criteria against which the cost-effectiveness of formal methods are measured in high-integrity computing differ from most other application fields. Where safety and security requirements are paramount and the clients are government sponsors, accountancy criteria become secondary, and a thorough documentation of the process is as important as any final product. If formal methods are seen to meet the needs of gaining assurance and of making the process visible, they can rightly be judged as having earned their keep. However, the criteria for mainstream clients differ in both respects. The formal methods community have little experience of framing available evidence of the benefits of formal approaches in the standard terms of cost, time, skills required, and products delivered.

Finally, even where there are detailed findings that formal methods can provide substantial productivity benefits, without necessarily requiring a massive re-training of the software labour force, these findings emanate almost exclusively from a community of highly-specialised experts who present the evidence to their peers. Conventional software developers, who have no foothold in the specialist networks of formal methodists may, indeed, find these particular messengers as peculiar as the messages they tell. The pattern of information trading, which works effectively within high-integrity computing, may solidify the idea that formal methods can only be effectively understood and used by those who are mathematically skilled already.

Thus, high-integrity computing shapes formal methods to fit contexts where conventional accountancy criteria are secondary to the search for greater assurance; where specialist (mathematical, rather than engineering) skills are already high; where links to academia are much closer than links to industry; where the documentation of the design process is as important to clients as the product itself; and where developmental rather than black-boxed tools are acceptable as output. In all these respects, the context of high-integrity computing encourages a shaping of formal methods into unusual configurations of skill, work organisation and technical achievement. These contingent factors seem likely to make it especially difficult for formal methodists to produce "objective" evidence that their techniques are potentially equipped to answer problems of productivity and cost regardless of the application field.

Finding the means to move formal methods of software development beyond their original laboratory of high-integrity computing may well prove difficult. In our sociological research at Edinburgh over the next few years, we hope to assess the issues sketched in this paper by means of detailed case studies on the genesis and use of varieties of formalisms.

References

[1] Coleman, D *The Technology Transfer of Formal Methods: What's Going Wrong?* Paper presented to the Workshop on Industrial Use of Formal Methods, Nice, March 1990

[2] Collins, H *Changing Order: Replication and Induction in Scientific Practice* Sage 1985

[3] Craigen, D (raporteur) *Government Forum* in *Formal Methods for Trustworthy Computer Systems Report from FM89* Dan Craigen (ed), Springer-Verlag 1990

[4] Friedman, A *Computer Systems Development: History, Organisation and Implementation* John Wiley & Sons 1989

[5] Hall, A *Seven Myths of Formal Methods* IEEE Software September 1990

[6] Harding S and Gilbert N *Taking Up Formal Methods* Paper presented to SPRU CICT Workshop on Policy Issues in Systems and Software Development July 1991, Brighton

[7] Ince, D and Andrews D *Software Engineering and Software Development* in Ince and Andrews (eds) *The Software Life Cycle* Butterworths 1990

[8] MacKenzie D *Inventing Accuracy: A Historical Sociology of Nuclear Missile Guidance* MIT Press 1990

[9] Macro, A *Software Engineering: Concepts and Management* Prentice Hall 1990

[10] Pelez, E *A Gift from Pandora's Box: The Software Crisis* PhD thesis, Edinburgh University 1988

[11] Quintas, P *Engineering Solutions to Software Problems: Some Institutional and Social Factors Shaping Change* Forthcoming in Technology Analysis and Strategic Management 1992

[12] Schon, DA (1963) *Champions for Radical New Inventions* Harvard Business Review Vol 41 pp. 77-86

[13] Sommerville, I *Software Engineering* Addison-Wesley 1989

[14] Tierney, M *The Evolution of Def Stan 00-55 and 00-56U* Edinburgh PICT Working Paper No 30 1991

3 The Social Negotiation of Proof: An Analysis and a further Prediction

Donald MacKenzie

3.1 Background

This paper is a preliminary report on one aspect of a research project studying the development of formal methods in computer science, which is being conducted as part of the Economic and Social Research Council's Programme on Information and Communication Technologies (PICT). The overall goal of the project is to discover the factors encouraging, retarding and shaping the emergence and adoption in the UK of formal methods. The approach being taken includes case-studies, historical work and international comparison.

The focus of this paper is a topic that may seem a strange one for social research: mathematical proof. Mathematical proof of the correctness (correspondence to specification) of software or hardware designs is being sought because of the certainty such proof is believed to provide, and which the empirical testing of systems is held not to provide. This paper is in no sense opposed to the search for proof: the increasing demand from regulatory bodies for proof is to be welcomed. Rather, the argument of the paper is that the pursuit of proof in computer system development involves the negotiation of what proof consists in.

The term "negotiation" is used here as a shorthand expression for a particular perspective on the development of knowledge. In this perspective, typical of the modern sociology of knowledge, it is argued that the application of concepts is determined neither by their past usage nor by any essential meaning that may be supposed to be inherent in them [1]. Their application in new situations is always potentially contestable. Nor is past usage sacrosanct: we may choose to see it as mistaken, and thus revise the "meaning" of the concept. This is the case, it has been argued, not only for the concepts of the everyday world, nor even just for those of empirical science, but even for the Platonist heartlands of mathematics and formal logic [2].

The metaphor of "negotiation" brings with it connotations of haggling, even of labour disputes and the smoke-filled rooms of the world of politics. These connotations are not accidental. Human interests are always involved in decisions as to how the network of concepts should be extended or revised [3]. In the case of scientific concepts, of course, these interests are often of an esoteric kind: preserving the reliability of certain kinds of prediction, maintaining the applicability of particular laws, and so on. Even here, however, the profane connotations of "negotiation" may not entirely mislead: the reputations and career interests of scientists are often tied up with these questions of the usage and extension of concepts, as an extensive literature of the sociology of science has shown [4].

The extension of the concept of proof from the world of mathematics and formal logic to that of computer systems critical to safety or security adds a further dimension to these connotations of "negotiation". Proof is becoming a regulatory and a commercial matter, as those with legal and administrative responsibilities start demanding it, and commercial suppliers start claiming to be able to provide it. As this happens, the social interests at stake broaden beyond scientific reputation and career, and the likely contexts of dispute shift from the scientific community to the marketplace and law court.

Four years ago, colleagues and I predicted that it might not be long before a "court of law has to decide what constitutes a mathematical proof procedure" [5]. Our basis for this prediction was the considerable variation, revealed by the history of mathematics, in the forms of argument that are taken as constituting proofs. The historian Judith Grabiner, for example, has shown how arguments that satisfied eighteenth century mathematicians were rejected as not constituting proofs by their nineteenth century successors such as Cauchy [6]. Our prediction rested on the assumption that attempts to prove the correctness of computer systems would bring to light similar disagreement about the nature of proof. This time, however, the actors in such dispute, and its context, would be quite different.

3.2 VIPER

That prediction has turned out to be correct, although, for contingent reasons, the legal case in question ended before a court ruled on the points at issue. The case concerned VIPER (Verifiable Integrated Processor for Enhanced Reliability), a microprocessor developed in the mid and late 1980s by a team of researchers from the UK Ministry of Defence's Royal Signals and Radar Establishment. Though VIPER has several other features designed to make it safe (such as simply stopping if it encounters an error state), what was crucial about it was the claimed existence of a mathematical proof of the correctness of its design. VIPER was marketed as "the first commercially available microprocessor with a proven correct design" [7].

The claim of proof became controversial. There has been sharp disagreement whether the chain of reasoning connecting VIPER's design to its specification can legitimately be called a "proof". In January 1991, Charter Technologies Ltd., a small English firm which licensed aspects of VIPER technology from the Ministry of Defence, began legal action against the Ministry in the High Court. Charter alleged, amongst other things, that the claim of proof was a misrepresentation, and sought damages under the 1967 Misrepresentation Act. The Ministry vigorously contested Charter's allegations.

Charter went into liquidation before the case could come to court. Nevertheless, the controversy surrounding VIPER, and the aborted litigation, reveal some of the scope for dispute over proof. The development of VIPER and the construction of its controversial proof are discussed elsewhere [8]. The core of the criticism of the claim of proof is as follows. The critics, Cambridge University computer scientist Avra Cohn, who worked on the proof, and Bishop Brock and Warren Hunt of the Austin, Texas, firm commisioned by NASA to evaluate it, use a definition of formal proof best summarized by Brock and Hunt's colleagues Robert Boyer and J. Strother Moore:

> A formal mathematical proof is a finite sequence of formulas, each element of which is either an axiom or the result of applying one of a fixed set of mechanical rules to previous formulas in the sequence.[9]

By that criterion, there is only a partial proof of the correctness of VIPER's design. This was constructed by Cohn, on contract to the Royal Signals and Radar Establishment, using HOL (Higher Order Logic), an automated system for proof construction developed by her colleague Mike Gordon. Even though large – her main proof consists of a sequence of over 7 million formulae – this work does not encompass all the steps between the top level specification of VIPER's behaviour and the logic-gate level description used to control the automated equipment employed to construct the "masks" needed to fabricate VIPER chips.

Cohn therefore wrote in 1989: "no formal proofs of Viper (to the author's knowledge) have thus far been obtained at or near the gate level" [10]. Brock and Hunt, likewise, concluded that "VIPER has been verified in the traditional hardware engineering sense, i.e. extensively simulated and informally checked" but not "formally verified" [11].

How would the claim of proof for VIPER have been defended, if the case had come to court? One can only speculate. The one published response (known to this author) by a member of the VIPER team to criticism of the claim of proof does not attempt a rebuttal [12], and, in any case, the defendant in the law suit was the Ministry of Defence, not the individual team members, so the line of argument adopted might, therefore, not necessarily have been theirs.

Nevertheless, it seems clear that a defence of the claim of proof would have had to involve challenging the notion of proof underpinning the criticism of it, so that mathematical arguments not conforming to the model summarized by Boyer and Moore could count as proofs. That, certainly, was the position adopted in defence of VIPER by Martyn Thomas, head of the software house Praxis, in an electronic bulletin-board comment on the end of the litigation:

> We must beware of having the term "proof" restricted to one, extremely formal, approach to verification. If proof can only mean axiomatic verification with theorem provers, most of mathematics is unproven and unprovable. The "social" processes of proof are good enough for engineers in other disciplines, good enough for mathematicians, and good enough for me ... If we reserve the word "proof" for the activities of the followers of Hilbert, we waste a useful word, and we are in danger of overselling the results of their activities [13].

3.3 Disputing "Proof"

These competing arguments, as they bear upon the VIPER proof, were never tested in law, and the shadow of litigation has inhibited open discussion of them in the scientific community; furthermore, there are several more specific issues that would have had to have been addressed in any assessment of the validity of the VIPER proof. However, the reference to mathematician David Hilbert in the quotation from Thomas points us to wider issues involved, and indicates that the VIPER case cannot be dismissed as simply the result of factors peculiar to this particular episode.

The concept of formal proof outlined by Boyer and Moore draws directly from the formalist tradition within mathematics spearheaded in the twentieth century by Hilbert. This tradition sought to break the connection between mathematical symbols and their physical or mental referents. Symbols were merely marks upon paper, devoid of intrinsic meaning. Proofs should be constructed by manipulating these symbols according to the rules of transformation of formal logic, rules which took a precise, "mechanical", form. For the formalist tradition, then, formal proof consists, in principle, of a sequence of formulae of precisely the kind described by Boyer and Moore.

The qualification "in principle" is, however, important. Formal proofs, in this sense, have been constructed for only relatively small areas of mathematics. Most proofs within mathematics do not take that form: they are shorter, more "high-level", more "informal". It is typically believed that they could be made fully formal, but the translation is seldom attempted. Part of the reason is the sheer tedium of producing formal proofs, and their length; this is also a large part of the attraction of automatic or semi-automatic proof generating systems, such as HOL or the theorem prover developed by Boyer and Moore.

The relatively informal nature of most mathematical proof could have been a resource for the defence of the claim of proof for VIPER, as we have seen in the quotation from Thomas. It was also the basis for the most influential general attack on formal verification of programs, a 1979 paper by Richard DeMillo, Richard Lipton and Alan Perlis. Proofs of theorems in mathematics and formal verifications of computer programs were radically different entities, they argued:

> A proof is not a beautiful abstract object with an independent existence. No mathematician grasps a proof, sits back, and sighs happily at the knowledge that he can now be certain of the truth of his theorem. He runs out into the hall and looks for someone to listen to it. He bursts into a colleague's office and commandeers the blackboard ... Mathematical proofs increase our confidence in the truth of mathematical statements only after they have been subjected to the social mechanisms of the mathematical community. These same mechanisms doom the so-called proofs of software, the long formal verifications that correspond, not to the working mathematical proof, but to the imaginary logical structure that the mathematician conjures up to describe his feeling of belief. Verifications cannot readily be read; a reader can flay himself through one of the shorter ones by dint of heroic effort, but that's not reading. Being unreadable and – literally – unspeakable, verifications cannot be internalized, transformed, generalized, used, connected to other disciplines, and eventually incorporated into a community consciousness. They cannot acquire credibility gradually, as a mathematical theorem does; one either believes them blindly, as a pure act of faith, or not at all [14].

Their paper provoked sharp criticism from defenders of the evolving practice of program verification. One wrote: "I am one of those 'classicists' who believe that a theorem either can or cannot be derived from a set of axioms. I don't believe that the correctness of a theorem is to be decided by a general election" [15]. Edsger Dijkstra, one of the leaders of the movement to mathematicize computer science, described the DeMillo, Lipton and Perlis paper as a "political pamphlet from the middle ages". However, his defence was of short, elegant, human (rather than machine) proofs of programs. He accepted that "communication between mathematicians is an essential ingredient of our mathematical culture" and conceded that "long formal proofs are unconvincing" [16].

Elsewhere, Dijkstra wrote:

> To the idea that proofs are so boring that we cannot rely upon them unless they are checked mechanically I have nearly philosophical objections, for I consider mathematical proofs as a reflection of my understanding and "understanding" is something we cannot delegate, either to another person or to a machine. [17]

At least three positions thus contended in the debate sparked in the late 1970s by DeMillo, Lipton and Perlis: the formal, mechanized verification of programs and hardware designs; the denial that verification confers certainty akin to that conferred by proof in mathematics; the advocacy of human, rather than machine, proof. Interestingly, the status of computer-generated proof was also the subject of controversy within mathematics in the late 1970s, controversy focusing on Appel and Haken's computer-based proof of the four-colour conjecture. The developers of this proof summarized both the objections and their defence thus:

> Most mathematicians who were educated prior to the development of fast computers tend not to think of the computer as a routine tool to be used in conjunction with other older and more theoretical tools in advancing mathematical knowledge. Thus they intuitively feel that if an argument contains parts that are not verifiable by hand calculation it is on rather insecure ground. There is a tendency to feel that the verification of computer results by independent computer programs is not as certain to be correct as independent hand checking of the proof of theorems proved in the standard way.
>
> This point of view is reasonable for those theorems whose proofs are of moderate length and highly theoretical. When proofs are long and highly computational, it may be argued that even when hand checking is possible, the probability of human error is considerably higher than that of machine error. [18]

3.4 Formal Proof and Rigorous Argument

No wholly definitive closure of the debates sparked by DeMillo, Lipton and Perlis, or by the proof of the four-colour conjecture, has been reached. The validity of the analogy between proofs in mathematics and formal verification of computer systems remains the subject of controversy. [19]

In the setting of standards for high-integrity computer systems, however, there are signs that the formal conception of proof will become dominant. Most interesting in this respect, because it addresses the issue directly, is the new UK

Interim Defence Standard 00-55, governing the procurement of safety critical software in defence equipment. It differentiates explicitly between "Formal Proof" and "Rigorous Argument":

> A Formal Proof is a strictly well-formed sequence of logical formulae such that each one is entailed from formulae appearing earlier in the sequence or as instances of axioms of the logical theory ...
>
> A Rigorous Argument is at the level of a mathematical argument in the scientific literature that will be subjected to peer review ... [20]

Unlike DeMillo, Lipton and Perlis, the Ministry elevates formal proof above rigorous argument:

> Creation of [formal] proofs will ... consume a considerable amount of the time of skilled staff. The Standard therefore also envisages a lower level of design assurance; this level is known as a Rigorous Argument. A Rigorous Argument is not a Formal Proof and is no substitute for it ... [20]

3.5   A Further Prediction

It remains uncertain to what degree software-industry practices will be influenced by Defence Standard 00-55: a procedure for granting exceptions to its stringent demands is embodied in the document. Nevertheless, similar standards for other sectors seem set to follow. While formal proofs of "real world" programs or hardware designs are still relatively rare, they will grow in number, in regulatory significance and in commercial importance.

The prediction of this paper is that as this happens, a new level of dispute and litigation will emerge. This will concern, not the overall status of computer-generated formal proofs (though that issue will surely be returned to), but an issue that has not hitherto sparked overt controversy: the internal structure of formal proofs. Even if all are agreed that proofs should consist of the manipulation of axioms according to "mechanical" rules of logic, it does not follow that all will agree on what these axioms or rules should be.

Again, the history of mathematics – and, in this case, the history of logic as well – reveals the scope for significant disagreement. The best-known dispute concerns the law of the excluded middle, which asserts that either a proposition or its negation must be true. At stake is thus the validity of mathematical proofs by reductio ad absurdum (showing that a proposition is true by proving the falsity of its negation). Formalists like Hilbert did not regard such proofs as problematic; while "constructivists" and "intuitionists", notably L.E.J. Brouwer, refused to employ them.

Another example is the Lewis principles, named after the logician Clarence Irving Lewis. These are that a contradiction implies any proposition, and that a tautology is implied by any proposition. They follow from intuitively appealing axiomatiziations of formal logic, yet have seemed to some to be dubious. In the words of one text:

> Different people react in different ways to the Lewis principles. For some they are welcome guests, whilst for others they are strange and suspect. For some, it is no more objectionable in logic to say that a [contradiction] implies all formulae than it is in arithmetic to say that x_0 always equals 1 ... For others, however, the Lewis principles are quite unacceptable because the antecedent formula may have "nothing to do with" the consequent formula [21].

Both excluded middle and the Lewis principles play their part in the practice of formal proof of computer systems. Excluded middle is widely used in automated theorem proof, for example in the HOL system used for the VIPER formal proof. The first Lewis principle – that a contradiction implies any proposition – is amongst the basic inference rules of the Vienna Development Method.

It would, however, be mistaken to conclude that future dispute over formal proof of computer systems will focus either on excluded middle or the Lewis principles. There is no reason to expect detailed replication of past controversies in mathematics or logic, although it is worth noting that the formalist/constructivist debate remains alive, and there has already been some intellectual skirmishing between the proponents of "classical" theorem provers, which employ the law of the excluded middle, and "constructivist" ones which do not [22].

Rather, the argument of this paper is more general. As formal proofs become of greater commercial and regulatory significance, powerful interests will develop in the defence of, and criticism of, particular proofs. Sometimes, at least, these interests will conflict. In such a situation, the internal structure of formal proofs will be subject to the most detailed, antagonistic, scrutiny. Dispute is then to be expected over the axioms and logical systems that underpin formal proofs, and that dispute will – precisely because of the new commercial and regulatory significance of mathematical proof – find its way into courts of law.

3.6   Conclusion

The internal structure of proofs is not substantively the most important issue in the application of mathematical proof to a computer system: on that criterion, it is outweighed by the issues of whether the formal specification of the system corresponds to what designers intend and whether the actual behaviour of a system will correspond to even the most detailed mathematical model of it [10]. Nevertheless, there is a sense in which the internal structure of computer system proofs raises what is intellectually the most tantalizing issue in the area.

In our culture, mathematical proof is the form of knowledge that is taken as having the best claim to the status of certain truth: given the premises, the conclusion must follow. Proof – especially formal proof – is the form of knowledge where it seems least likely that we will observe the phenomena of negotiation: the clash of contending interests and points of view, and their resolution not by the unequivocal application of absolute standards of truth or rationality but through more mundane processes such as debate within scientific communities or judgements by courts of law.

Formal proof of computer system correctness is, therefore, of interest because it pits one of our strongest common-sense intuitions about the nature of knowledge against a fundamental principle of the modern sociology of knowledge: that all development of knowledge, however apparently absolute its grounds, potentially involves processes of negotiation. That principle is the ultimate basis of the prediction of dispute and litigation over the internal structure of formal proofs. As computer system proof grows in significance and moves into the commercial and regulatory worlds, we will have a chance to see whether our ordinary intuitions about mathematical proof, or the sociology of knowledge, is correct.

References

[1] Barnes B. *T.S. Kuhn and Social Science.* Macmillan, London, 1982

[2] Bloor, D *Knowledge and Social Imagery.* Routledge and Kegan Paul, London, 1976

[3] Barnes, B. *Interests and the Growth of Knowledge.* Routledge and Kegan Paul, London, 1977

[4] Shapin, S. *History of Science and its Sociological Reconstructions.* History of Science 1982; 20: 157-211

[5] Pelaez E, Fleck J, MacKenzie D. *Social Research on Software.* Paper read to National Workshop of Programme in Information and Communications Technologies, Manchester, 1987

[6] Grabiner JV. *Is Mathematical Truth Time-Dependent.* American Mathematical Monthly 1974; 81: 354-65

[7] Hughes NH: Foreword to Charter Technologies Ltd., VIPER Microprocessor Development Tools, December 1987

[8] MacKenzie D. *The Fangs of the VIPER.* Nature 1991; 352: 467-68

[9] Boyer RS, Moore JS. *Proof Checking the RSA Public Key Encryption Algorithm.* Am Math Monthly 1984; 91: 181-89

[10] Cohn A. *The Notion of Proof in Hardware Verification.* J Aut Reasoning 1989; 5: 127-139

[11] Brock B, Hunt WA Jr. *Report on the Formal Specification and Partial Verification of the VIPER Microprocessor.* Computational Logic, Inc., Austin, Texas, 15 January 1990 (Technical report no. 46).

[12] Kershaw J: Foreword to [11].

[13] Thomas M. *VIPER Lawsuit withdrawn.* Electronic mail communication, 5 June 1991

[14] de Millo RA, Lipton RJ, Perlis AJ. *Social Processes and Proofs of Theorems and Programs*. Comm ACM 1979; 22: 271-80

[15] Lamport L. As quoted in Shapiro SS, *Computer Software as Technology: An Examination of Technological Development*. PhD thesis, Carnegie Mellon University, 1990

[16] Dijkstra EW. *On a Political Pamphlet from the Middle Ages*. ACM SIG-SOFT, Software Eng Notes April 1978; 3(2): 14-16

[17] Dijkstra EW. *Formal Techniques and Sizeable Programs*. In Dijkstra, *Selected Writings on Computing*. Springer, New York, 1982, pp. 205-214

[18] Appel K, Haken W. As quoted in Davis PJ, Hersh R, *Why should I believe a Computer?* In Davis, Hersh, *The Mathematical Experience*. Harvester, Brighton, 1981, pp. 380-87

[19] Fetzer JH. *Program Verification: The Very Idea*. Comm ACM 1988; 31: 1048-63

[20] Ministry of Defence. *The Procurement of Safety Critical Software in Defence Equipment*. Ministry of Defence Directorate of Standardisation, Glasgow, 5 April 1991 (Interim Defence Standard 00-55/Issue 1)

[21] Makinson DC. *Topics in Modern Logic*. Methuen, London, 1973

[22] Manna Z, Waldinger R. *Constructive Logic considered Obstructive*. Typescript, no date

4 On constructing large software systems

Jean-Raymond Abrial[1]

4.1 Introduction

Although we have in mind a particular approach of software construction which we have been developing and using for some years, our intention is not to describe a specific method here, it is only to give a number of guidelines and rules of thumb that we re-discover little by little and which can be used, we think, whatever the technical environment at one's disposal. Our main concern is the following:

How can we avoid being overwhelmed by *complexity* ?

Our belief is that part of the complexity we face is *not* due to the problem itself: it is due to ourselves. So, the complexity we shall try to reduce is the one we can, in principle, control. If the problem to solve is complex by itself, the last thing we want to do is to introduce more complexity. We are not going to present any new or grandiose principles of organization. The best way, we think, to fight complexity is

By introducing ... *simplicity*.

The classical argument against this kind of slogan is well known: this is appealing but, at the same time, it is also very naive. Let's be naive then and let's explain why.

4.2 People

One should never forget that computer systems are built by human beings. In this section, we survey some of the difficulties that are due to this fact.

Programming is already a very difficult, demanding and intense cerebral activity, so much so, we believe, that a program always reflects the *mental structure* of its author. In the programming of large software systems, the difficulty is magnified by the very fact that a large number of people are involved. By analogy with what we have just said of the single programmer program, it is our belief that a software system reflects the *social structure* of the team that built it. Over the last decades, people have been speaking of the ever growing *software crisis*. We do not believe in *software crisis* only in *social crisis*.

Shifting requirements, staff discontinuity, difficulties defining clear boundaries, management problems, all these, and many others, are common difficulties encountered in the construction of complex systems. Unlike other technical activities however, software construction is *extremely sensitive* to these problems which are all reflected and amplified in the final product.

Among the previously mentioned problems, the hardest, we believe, is the management problem: most software projects are *ill managed*. This is so because the classical management practice (i.e. report, inspection, measurement, deadlines) is simply not adapted to the complexity of the programming task.

---

[1] With the kind permission of ELSEVIER, this paper is a reprint of a paper published in the IFIP 1992 proceedings

Unlike other technical activities, software systems are almost exclusively written by *junior people* facing the incomprehension of their senior managers. The latter simply cannot understand what their staff do because of the high level of complexity; and the *response* of the former is, of course, an increase of complexity. What is at fault here is the classical dogma that claims that managers need not be actively involved in technical tasks precisely because their job is to control it.

We believe, on the contrary, that the manager of a software project has to be intensely technically involved in the project: its role is to ensure, from the inside, that no social disease develops, it should not be a role that, in itself, is a social disease. Rather than controlling people, he has to ensure that proper communication links are established between members of the team. When people do communicate harmoniously, their programs have a certain tendency to do so too.

Rather than being the person who puts pressure on the team to meet deadlines, the manager of a large software project should be prepared to take the *negative* decision to throw away what has been done so far and to redo it from scratch. He will discover, to his own astonishment, that he might very well, as a result, meet some later deadlines.

A complex software system that works satisfactorily is, quite often, one that has been re-written at least three times *before* its first final delivery.

Of course, such a practice might seem to be very strange to the industrial milieu: this is so because our mental model of industrial objects is that of *material* objects not that of *soft* artefacts where the question of the final effective construction has disappeared. It is our belief that no previous model of work organization is valid for software construction: complexity and the particular activity of programming require some new models which have not yet been discovered. In particular, the presence of worldwide public networks might change completely the picture. People need not be physically concentrated in the same place: curiously enough, that separation might improve their working relationships.

Clearly, nobody could ever enter into the finest details of a complex system. This is where *abstraction* and *visible interfaces* play a fundamental role. People need only to have the possibility to participate in some technical "rendez-vous" where what is discussed is clear to everybody.

4.3   Frames

Framing is the answer to questions such as: What to specify? What to document? What to make a design of? What to translate into final code?

A number of programming features have been proposed, for quite a long time, to address that sort of question. Such features are known under various names such as: *Class, Module, Package, Abstract Data Type, Object*. Such techniques, however, were only designed with programming in mind. The idea is to enlarge the usage of such concepts to the *complete process* that goes from specification and design down to the final code. For this, we shall use the somewhat neutral term of *frame*. In what follows, we shall study some aspects of the enlargement of this concept.

Each frame specification constitutes the *mathematical model* of a future *programming module*. Whatever its precise concrete syntactic structure, the *specification* of a frame should be made of the following components:

- a number of set-theoretically typed variables,

- a number of invariant conditions on these variables,

- an initialization,

- a number of parameterized operations,

- a number of pre-conditions on each of these operations.

The variables, together with their types and invariant, form the *proper state* of the frame. The initialization expresses some desired properties of the initial state of the frame. Each parameterized operation expresses some desired properties of the frame evolution and communication with its environment. These elements are not independent: the initialization should make the frame variables forming a proper state, and each operation, provided it is subjected to its pre-conditions, may only change the frame variables from proper states to proper states. The state invariant and the operation pre-conditions are predicates formally written using a set-theoretic notation as described in section 4.

Operations, whatever their precise syntactic structure, have input and output formal parameters and also a body which can be built by means of various *pseudo-programming statements* such as:

- multiple assignment of variables,

- choice between two or more statements,

- multiple execution of two or more statements (dealing with independent variables),

- non-deterministic introduction of typed variables in a statement.

Of course, some classical *programming statements* such as skip (the statement that does nothing) or the conditional statement can also be used in frame specifications. This is not the same, however, as sequencing or loop which are clearly *inappropriate* in frame specifications since they are far too precise in describing the way algorithms are eventually implemented. All such programming statements will be summarized in section 10.5.

Frames can be subjected to a number of transformations which, whatever their precise syntactic formulation, are among the following:

- inclusion,

- refinement,

- abstraction,

- importation.

The first transformation, inclusion, allows you to build a large frame specification from smaller, more manageable, ones. The latter are said to be *included* in the former.

The second transformation, refinement, is the process by which a frame is given a *more concrete version*. This is done by replacing some of the variables of the former by more concrete ones, by possibly weakening some of the pre-conditions of its operations and, finally, by replacing some of the pseudo-programming statements of its operations by genuine programming statements or, at least, by more deterministic statements. Refinement usually requires several steps thus transforming little by little a frame specification into a programming module. Notice that the name and parameters of the operations of a frame remains the same through refinement so that the *external appearance* of the frame is unchanged by the various refinement steps it may be subjected to. Of course, a refinement, in order to be correct, must be such that the new version of each operation preserves the semantics of its corresponding more abstract formulation. From the point of view of the practitioner, refining a frame consists in *re-writing* it in a more concrete (implementable) fashion.

The third transformation, abstraction, is the converse of the previous one. One might wonder, at first glance, why we need at all such a frame transformation. This is so because experience shows that the choice of the right level of abstraction is a very difficult task. When you eventually discover that the level of abstraction of a certain frame is too low (a classical mistake of beginners) then you had better abstract out from it, thus making a more general version which is less specific.

The fourth operation, importation, is the one by which the final refinement of a frame can invoke the specification of other frames. The latter are said to be *imported* in the former: they act as "abstract machines" on which the final refinement is implemented. The imported frame specifications are referenced in the importing one by means of *calls* to the relevant operations only. Since, as we have noticed previously, frame operation names and parameters are not changed by refinements, such imported specifications will be replaceable later by their corresponding *final refinement*, and so on. In this way, we construct, again little by little, the overall layered structure of our eventual software system.

Frames should be developed with re-usability in mind (this is where the possibility of abstracting a frame might play a central role). This concern is definitely a social one. A team that works with a long term goal might have the natural tendency to invest in the future by thinking ahead.

For instance, (parameterized) frames could be written *once and for all* that handle data bases of mathematical objects such as finite sequences, sets, trees, binary relations and so on. Operations will thus be provided to create, remove, modify and access such objects independently from their eventual implementation. And, we may think having different refinements of these data base specifications to work in different contexts ranging from reactive systems to large information systems.

4.4 Sets

In order to formally describe the specifications and refinements of frames as introduced in section 10.2, we believe that there is no need to develop a particular specification language of any sort. It is simpler, safer, and more understandable to use *directly* the mathematical notation of set theory. In this section, we survey the basic fundamental concepts of set theory that are useful for us and the way they can be introduced rigorously.

One of the problems with set theory is that, over the last hundred years, it has become *too complicated*. This complexity has frightened people. For instance, there is no need, for our purposes, to use the full technical details of the Zermelo-Fraenkel set theory. In fact the original Zermelo set theory augmented with a few more concepts will fulfill completely our needs. One immediate advantage of this decision is that every set-theoretic expression can be *typed* using a simple *proof procedure*. The only basic constructs we need are the most obvious ones, namely:

- cartesian product,

- power set,

- set comprehension.

The axiomatization of set theory is then done in a straightforward manner by defining the form taken by *set membership* on these various constructs. The *equality* of two sets is expressed, of course, by the identity of their memberships. From there, it is easy to extend the notation by introducing some *conditional definitions* such as those of:

- the classical operations on sets (inclusion, union, intersection, ...),

- the empty set,

- the possibility of defining a set by listing its elements,

- the binary relations and their related operations,

- the functions and their related operations.

In order to construct formally more elaborate *mathematical objects* such as natural numbers, finite sequences, finite trees and their related operations, we only have to postulate the existence of an infinite set and that of a choice operator on set. Note that the concept of *finiteness* can be defined rigorously by means of the previous basic concepts and definitions.

4.5 Programs

Our view on programming languages is *extremely conservative*. It is our very strong belief that programming languages are far too complicated. In this section, we present what we think should be the programming notation to be used for programming large software systems. What is presented here is *not a new language proposal*: it is nothing else but that part which is *common* to every imperative programming language. As a side effect, the question of portability, typically a self-inflicted wound, would thus disappear as an issue.

A final complete software system is made of a number of *modules*. Each module is made of the following components:

- a number of typed variables,

- an initialization,

- a number of parameterized operations.

A module A may *import* one (or several) module(s) B, meaning that within the operations of A you may call operations of B but not refer to the variables of B.

The typing system for variables is more than simplistic: variables can only be either of type *integer* (a pre-defined finite interval of natural numbers) or of type *array* (total functions from a finite interval of *integers* to an *integer* or to another *array*). As you may have noticed, no *record* and no *pointer* (since the corresponding functionalities can be fulfilled by other means).

An operation may have input and output parameters (mind the plural) supposed to be of type *integer* only. Operations, whatever their precise syntactic structure, can be built by means of various *programming statements* such as:

- skip (the statement that does nothing),

- multiple assignment of integer variables or scalar array elements,

- conditional statements,

- local introduction of variables in a statement,

- sequencing,

- loop,

- operation call.

Local variables are of type *integer* only. Arithmetic expressions may be formed, as usual, with the basic biased arithmetic operators. Boolean expressions (which are needed in conditional or in loop statements) are formed, as usual, with the basic relational and boolean operators.

As you can see, this programming notation appears to be a *sub-notation* of the one introduced in section 2 for frame specification and refinement. In fact, a module is nothing else but a certain *frame refinement*. We have thus a *notational continuum* between specifications, refinements and programming modules.

To the people who claim that this view of programming is so naive and oversimplified that nobody could ever use it in a real project, we just ask to have a look at the source code of any real system: 95% of it, at least, is just that. For the remaining 5%, use any escape you want through some specialized library.

4.6   Proof

It is our belief that the overall construction process of a complex software system and the proof of its correctness *are the same thing*. The mechanisms by which mathematicians make a complex proof comprehensible to themselves and to their colleagues are of the same nature as those by which software designers make a system comprehensible to themselves and to their colleagues: abstraction, parameterization, reformulation in another variable space, structuring, re-use of available elements.

As we have seen in previous sections, the proofs that accompany the construction of a frame are of the following nature:

- the initialization of the frame specification should establish the invariant,

- the operations of the frame specification should preserve the invariant,

- the initialization and operations of a frame refinement should preserve the semantics of their corresponding more abstract versions.

Such proof obligations can be converted into corresponding formal statements of set theory and first order predicate calculus. It is our experience that difficulties encountered in proving such statements might sometimes disappear by re-thinking the way a frame is specified and refined. The complexity of the proofs thus forces us to simplify the design of the system. Most of the time, however, such proofs are not difficult; the only problem is in their number; yet another complexity that might overwhelm us: tools are needed to help us here.

From the formal construction of set theory alluded above in section 4, it is possible to derive a number of mathematical laws that can be later used to prove the previously mentioned formal statements. Among these, the algebraic laws of the classical set operations are already well known. Other similar and equally simple laws could be derived concerning relations and functions thus forming a very powerful relational calculus. Finally, the mathematical objects (natural numbers, sequences, trees) mentioned in section 4 also have well known properties that could be derived and listed. Moreover, their very definitions lead naturally to powerful induction rules.

It is not claimed, of course, that we could ever reach a point where a complete set of rules will be eventually available. Our belief, however, is that, little by little, we will reach a situation where most (say 80%) of the proof obligations generated by the systematic construction of a software system could be discharged blindly by means of such rules. For the remaining proofs, a human intervention will always be needed.

4.7   Tools

In view of the concept of frame that we have introduced in section 2, we now consider how the classical process of *compilation* has to be generalized. A frame is not just represented, as was a programming module, by a pair of "source-object" codes. We now have a *sequence of formal documents* ranging from the *specification* of the frame to its final *implementation* through a series of intermediate *refinement* documents. At any such stage, a frame is subjected to a number of processings which are essentially of the following nature:

- type-checking,

- proof obligation generation,

- proofs,

- translations,

- animation,

- automatic generation.

Type-checking is the classical procedure by which it is verified that certain straightforward semantic laws are obeyed. The type-checking we may have for frames is essentially of the same nature as the one encountered in classical compilers.

Proof obligations, as we have already presented them in the previous section, have to be generated automatically. Proofs have to be done eventually by a mixture of automated and interactive provers.

Translations are only relevant for the last refinement of a frame. They are of the same nature (although being simpler) as those encountered in classical compilers.

Animation is only relevant for the specification of a frame. The idea is to have a tool able to execute the operation of a frame specification (as much as it can, of course). This might be a good way to figure out whether a given frame specification corresponds to our informal needs.

Finally, as we have explained at the end of section 3, it might be useful to have at one's disposal some ready-made (and proved) frame specifications and corresponding final implementations that are related to well understood areas. As each such frame, obviously, would belong to some larger families, we could envisage the construction of specialized tools allowing us to generate frames automatically on demand.

Clearly, this overall approach does generate a *very large number* of formal documents. As a consequence, a tool ensuring that these documents are correctly handled is absolutely indispensable. Among others, such a tool should have the well known capability to re-make a number of operations done by other tools on a frame that has just been modified and, of course, to determine the influence of such modifications on the status of other frames. Another useful capability is one by which the frame population could be queried.

4.8 Conclusion

Although our initial claim was to introduce simplicity, it is not clear whether we have indeed achieved our goal. The "simplicities" we have introduced may appear, after all, to be quite sophisticated *once put together*.

Whatever the techniques and the tools at one's disposal, the proper construction of a large and complex software system will always remain a very difficult problem. In this area, we do not believe in ready-made solutions.

And the key to success will depend, eventually, on the spirit of the people involved.

4.9 Acknowledgments

What has been presented here is borrowed from many people. The influence of the ideas of E.W. Dijkstra, C.A.R. Hoare, C.B. Jones and D.L. Parnas is clear. Specification and refinement techniques, as introduced by C. Morgan, He Ji Feng, K.A. Robinson, D. Gries and others also provide a rich vein. The project that provides some of the ideas presented here has been financed over the years by **British Petroleum** and by **GEC-Alsthom**. With these companies, I have had the great pleasure to work with teams headed by I.H. Sørensen and F. Mejia respectively.

5 Composition of Descriptions : A Progress Report

Pamela Zave
Michael Jackson

5.1 Introduction

We have embarked upon an ambitious research project, and achieved some encouraging early results. These results will be presented in two forthcoming papers.

This progress report establishes a context for the results. It explains our overall goals and motivations, imposes a loose organizational structure on our work, and shows where some tasks (accomplished, in progress, and postponed) fit in.

5.2 Why Compose Descriptions?

We are investigating the possibility of developing systems by composing descriptions of all the properties they should have. Two familiar examples illustrate this concept.

In CSP [6] a process can be regarded as describing a behavior set, a characteristic property of the behavior set, or an executable process capable of exactly the behaviors in the set. The concurrent composition of a set of CSP processes describes a behavior set that satisfies the properties of all composed processes simultaneously. A system capable of exactly the behaviors in this set can be implemented by performing a dynamic (runtime) composition, based on shared events, of executable versions of the contributing processes.

In JSP [9] the control structure of a sequential program is obtained by composing a regular expression describing the structure of its input data with a regular expression describing the structure of its output data. The regular expressions are depicted as trees, and the composition of two trees is the smallest tree having both as subtrees. Unlike the dynamic CSP composition of our previous example, composition of JSP trees is usually performed statically, and yields a control skeleton for a program in some target language.

We have four major reasons for being interested in composition of descriptions as an approach to system development:

1. Composition has potential for increasing automation: people would describe desired properties, while computers would check their consistency and generate software. Methods based on composition would shrink giant steps such as design, and therefore minimize the many problems of reproducibility, traceability, and verification such giant steps entail.

   An approach based on prototypes and transformational implementation (automated optimization), with a similar goal of increasing automation, has been proposed [1, 18]. Transformational implementation is difficult, particularly when a series of transformations is contemplated, because a prototype embodies many implementation decisions that must be unmade before they can be remade. We are interested in knowing if it is possible to avoid this difficulty by allowing incomplete descriptions until the very final development step, and never forcing a premature decision.

2. Composition has potential for increasing reuse. Today few artifacts of the software-development process are reusable, because each artifact is the result of intertwining many concerns, decisions, and properties. In another situation nine out of ten of those decisions may still be appropriate, but if the tenth one is not appropriate—and cannot be separated from the nine good ones—then the artifact is useless. The ability to compose implies an ability to decompose, to separate concerns, so that each relevant concern can be described separately. Some properties are required of many systems, so their descriptions should be highly reusable. Each development project should be able to select and compose these descriptions as needed.

A sort program, for example, can be specified in terms of an unsorted and a sorted sequence such that the two sequences have the same elements, and in the sorted sequence the representational order is the same as some logical order on the elements. Many other concerns are separable from this definition, including the type of the elements in the sequences, the logical order on the elements, the representational orders of the sorted and unsorted sequences (Temporal or spatial? If spatial, is access random or sequential?), which sequences form corresponding unsorted/sorted pairs, which of a pair is considered the input and which is considered the output, and any resource constraints. Programming-language design has been making steady progress in decomposing and composing such concerns, by means of features such as polymorphism, higher-order functions, and unification. Yet programming languages must satisfy severe and diverse constraints, and we wonder if, in an experimental context with fewer constraints, a simpler and more general mechanism can be found.

3. A sufficiently general composition mechanism would allow developers to use many formal languages, describing each property in the language best suited to expressing it clearly and concisely.

The need for multiparadigm description of system properties is well known, but the issue of analysis is often neglected. Each formal language is the result of a design trade-off between power of expression and susceptibility to algorithmic manipulation (consistency checking, verification, analysis, transformation, and execution). Thus one of the most important reasons for multiparadigm description is to express each set of related properties in a language whose accompanying tools can analyze them.

The majority of multiparadigm research projects are aimed at designing new programming languages integrating the features of many paradigms (Hailpern provides a representative sample [4]). While this is a practical approach to supporting multiparadigm programming in the near future, it is too inflexible for our goals. Considerations of cost and complexity will limit it to combining particular representatives of a few of the most popular paradigms, while we are interested in experimenting with a wide variety of notations, including those that are highly specialized or similar to each other. Furthermore, merging languages tends to compromise their analyzability, as most algorithmic manipulations of formal languages are quite sensitive to the features of the languages.

4. Composition may provide a means for understanding and comparing development methods. According to our viewpoint all methods compose descriptions, but the compositions are usually predetermined and implicit. If composition of descriptions is indeed a fundamental software-development step, then it should be possible to explain most methods in terms of explicit descriptions and compositions, and to compare or combine them on the basis of these explanations.

Current investigation of the software process concentrates on the relationship between the descriptions that must be produced and the tasks (and coordination) needed to produce them. We are interested in understanding the relationship between the problem that must be solved and the descriptions needed to solve it. These perspectives are very different, and may complement one another.

5.3 What Is Described?

Our goal will be impossible to achieve unless we are are very careful about what is described. The problem to be solved and the desired properties of the solution must be described completely. Distinct concerns must be kept separate, and not confused or conflated.

The primary distinction, made in all development methods, is between the *domain* and the *system*. A domain is a fragment of the real world (whether actual or anticipated, tangible or intangible) requiring automated support or control which the system is intended to provide.[2]

Two kinds of description are concerned exclusively with the domain. The first describes relationships that are true of the domain whether the system exists or not. Many facts about the domain are relevant to a system-development project—usually many more than are recorded [8]. The current interest in "domain modelling" [14] and "domain knowledge" [2] emphasizes the need for descriptions of this kind.

The second kind of domain description concerns relationships that the system is intended to enforce, *i.e.*, requirements. Most specification methods purport to describe a system and its interface to its environment (Lamport explains this viewpoint clearly [15]). Our perspective is different because we interpret requirements as descriptions of the domain rather than of the system.

A trivial example illustrates the difference. It concerns a banking system for an individual depositor. The depositor will initiate transactions, each of which has a monetary argument (positive for a deposit, negative for a withdrawal).

---

[2] "Environment" is often used instead of "domain," on the grounds that a system is embedded in its environment and the environment is the complement of the system. We prefer "domain" because the word emphasizes the more active and substantive role we think domains should play.

The system will respond to each transaction with an acknowledgment[3] whose argument is the depositor's account balance after the transaction.

The usual solution to this problem specifies a system in which input events are transactions, output events are acknowledgments, and state is the current account balance. While we might accept a syntactically similar or identical description, we would interpret it as describing a desired relationship among transaction events, acknowledgment events, and account balances, *all of which are observable phenomena of the domain.*

This subtle difference has many implications, because implementation bias is a chronic problem in specifications of systems. Suppose that, in a system specification, it is necessary to represent a collection of distinct elements. Should the representation be a set or a sequence? If the particular functions being specified require ordered elements then a sequence must be used; if the functions do not require an order then only a set representation avoids implementation bias [13].

The trouble with this line of reasoning is that it may be difficult to decide exactly which representation decisions are justified by the required functions. Furthermore, it may force the developer to omit an element order that is relevant to the problem and will be required by the very next function added to the system. Thus needs for maintainability and freedom from implementation bias are brought into conflict.

Another symptom of the underlying problem is the perennial ambivalence about specifications that mention the internal state of the system. Many specification methods do so, of course, but there is enough suspicion that internal state constitutes implementation bias (because the internal state of the system is not visible at the system/environment interface) to motivate different approaches. The presence or absence of internal state is one of the main characteristics distinguishing model-oriented from property-oriented methods [16]. Algebraic and purely temporal methods avoid mentioning internal state. The defense of state in the transition-axiom method [15] is based partly on the particular semantics of a transition-axiom specification (in which an automaton is formally equivalent to the grammar that generates the language accepted by the automaton), and partly on the convenience of using state in system specifications.

Describing requirements in terms of the domain rather than the system avoids all such issues. A description is valid if and only if it is true (or required) of the domain, regardless of whether it is absolutely necessary to the development project at hand. Superfluous descriptions can be detected. Descriptions of the state of the domain are welcome. Descriptions of the system concern only obvious implementation issues such as properties of computing machines (made of hardware and software) and properties of connections between machines and the domain.

In addition to the above distinctions, many development problems are more easily solved by describing them in terms of several domains and several systems (a paper by Jackson on composition of descriptions [10] contains many examples). For the time being, however, we are concentrating on domain descriptions in simple development problems, each with a single domain and a single system.

---

[3] In this trivial example, transactions never fail. The account balance can be negative.

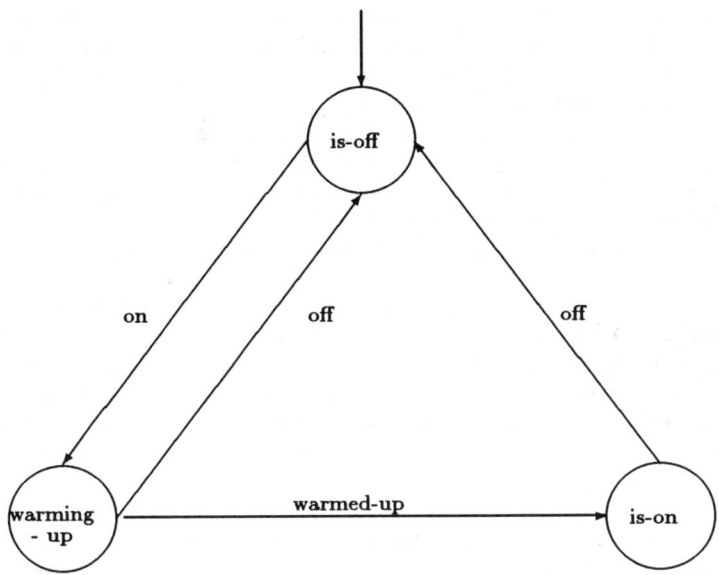

Figure 1: An FSA describing a fluorescent light.

5.4 What Is a Description?

We have identified several examples in which the meaning of a formal specification as a mathematical object is clear, but its meaning as a description of a real-world domain (or even a system to be developed) is unclear [12]. In the context of today's system development this is not usually a problem: vagueness of the specification method is tolerable in aspects of the system that are clearly understood and developed manually. But in the context of our goal such ambiguity is unacceptable. An automatically generated composition of many descriptions will be meaningless garbage unless the content and role of each contributing description is very precisely defined.

The deterministic finite-state automaton (FSA) in Figure 1, describing a fluorescent light, illustrates some of the ambiguities. Its meaning may seem plain enough, but actually depends on which domain phenomena the FSA is describing. Depending primarily on its scope—the set of domain phenomena constrained—the FSA has at least four possible interpretations:

1. If its scope is the event types *on, off,* and *warmed-up* then it describes a set of possible event sequences over that alphabet. It is formally equivalent to a regular expression, regular grammar, or unreduced FSA describing the same language. This is the interpretation used in the transition-axiom method.

2. If its scope is the states *is-off, is-on,* and *warming-up* then it makes cer-

tain assertions about the state of the domain. These assertions include that the domain is always in exactly one of these states, and that the domain can never go directly from the *is-off* state to the *is-on* state without passing through the *warming-up* state. The description says nothing about events. This interpretation, like the following two others, is *not* equivalent to a regular expression, regular grammar, or unreduced FSA. In these interpretations the states of the FSA are direct descriptions of the states of the domain, so the FSA cannot possibly mean the same thing as a description with different states or no explicit states.

3. Its scope may include both states and event types. In this interpretation the FSA makes the same state assertions as (2), and also asserts that, for every transition, if the domain is in the source state and an event of the type labeling the transition occurs, then the domain leaves the source state and enters the destination state. This interpretation does not constrain events in any way; it is the interpretation used by Statecharts [5].

4. The final interpretation has the semantics of (3), with the additional constraint that if the domain is in a state with no out-transition labeled by a particular event type, then events of that type cannot occur. It thus combines the semantics of all three previous interpretations.

In a forthcoming paper [12] we show how to represent domain phenomena formally and we define domain descriptions. Our definition has the precision to distinguish between all four interpretations of the FSA, and yet the flexibility to allow all four of them to be used. We show how descriptions can build on one another (in the same sense that mathematical theories build on one another), and that these ideas are applicable to a wide variety of description languages.

One major limitation of our current results is that we use only behavioral semantics: what behavior could be recorded by an alien observer of the domain, capable of recognizing certain phenomena but incapable of guessing causality or who is making the choices. In other words, we have deferred treatment of *control*.

Perhaps the best way to explain what is missing is to give some examples of its presence. In the transition-axiom method each action is associated with an agent, which is either the system or some aspect of the system's environment. Whenever an action is logically enabled, its agent has the unilateral power to make it occur. In CSP the trace semantics (which corresponds to our behavioral semantics) must be augmented to capture control features such as internal choices that affect the external behavior of processes.

Obviously we think that behavioral and control issues are somewhat separable, or we would not have organized our work in this fashion, but control issues are ubiquitous and fundamental. For example, many consistency conditions are dependent on control. Consider two event languages L_1 and L_2 over an alphabet of input events to a required system, such that L_1 is properly contained in L_2. If L_1 is a description of all the input event sequences that the domain can possibly generate, and L_2 is a description of all the input event sequences that the system can parse and respond to without crashing, then the descriptions are consistent. If the roles of L_1 and L_2 are reversed then they are not consistent; the difference in the two cases arises because the domain can control what input events occur, and the system cannot.

5.5 What Is Composition?

The most pressing need is for a definition of consistency among descriptions, and for a definition of the semantics of the composition of a set of consistent descriptions. Eventually we need to understand how to perform explicit compositions, whether dynamic or static, but that issue can be deferred until the meaning and usage of composition is clear.

The meaning of an individual domain description is an assertion over domain phenomena. Obviously the assertional meanings of all description languages should be given in a common semantic framework, so that the meaning of composition is conjunction of assertions within the framework.

The success of conjunction as composition depends on the choice of common semantics. The following properties are important for achieving our goal:

1. The common semantics should be compatible with the formal representation of domain phenomena mentioned in Section 8.4.

2. The common semantics should accommodate a wide variety of description languages, including all of those in common use for specification, and also including popular notations (such as dataflow diagrams, decision tables, queueing networks, and Gantt charts) usually considered too informal or too specialized for purposes of formal specification.

3. For the sake of familiarity and reasoning capabilities, the common semantics should be based on a simple logic. Gordon lists a set of logics in order of ascending expressive power [3]. Classical first-order predicate logic with equality and induction principles is low in this order, and seems sufficient at this time.

4. It should be possible to compose any consistent set of descriptions, regardless of overlaps or gaps in scope, regardless of which languages they represent, and regardless of where boundaries between languages are drawn. This contrasts with many *ad hoc* techniques for composition, which rely on strict assumptions about the languages used and the properties specified in each. The most common example of this is the control/data partition; it has been proposed in numerous variations, including recently the LOTOS/Act One partnership [7]. Although such rigid schemes make composition possible and may have other advantages (see below), our goal demands more flexibility.

In a forthcoming paper [11] we present a common semantics having all four properties. The paper emphasizes how an innovative encoding supports property (4), which is by far the most difficult one to achieve. It also gives many examples of composition of descriptions.

In the absence of information about control, consistency of a set of descriptions is defined as satisfiability of their conjunction. This is a necessary but not sufficient condition for full consistency.

It should be noted that we regard correctness conditions as normal descriptions (of required domain relationships), rather than as artifacts of a different kind. From this perspective consistency subsumes verification, and consistency checking is the primary goal of formal reasoning.

The only general approach to consistency checking would be to translate all partial specifications into the common semantics, and attempt to determine satisfiability (and other properties) within the predicate calculus. Such an approach would be absurdly inefficient. It is clear to us that practical consistency checking must exploit the characteristics of the particular specification, just as a user of a high-level language exploits its particular features to specify properties concisely and comprehensibly.

Practical consistency checking will depend on the analytic capabilities of the languages used, on the assignment of properties to be described to languages for describing them (the paradigm decomposition), and on the degree of redundancy among partial specifications. This should be a fruitful area of research, as the issues involved are extremely interesting;[4] one example is the redundancy trade-off. The less redundancy there is between partial specifications, the easier it will be to establish their consistency, because only the overlapping parts of the specifications need be checked. Yet sometimes the scope of a description must be expanded—introducing redundancy with other descriptions—so that the description will be self-contained with respect to certain properties analyzable within that description's toolkit [17].

We have not attacked this problem directly yet, but we have made an indispensable contribution to its solution in defining a composition mechanism that gives an intuitive formal semantics to a multiparadigm specification regardless of the paradigm decomposition used. In other words, we have provided freedom to design paradigm decompositions in the best interests of consistency checking.

5.6 Description Reuse

We have been mindful of the need to enhance the reusability of descriptions, and have achieved modest success with a simple renaming operation on descriptions. But this is a complex area of investigation in its own right, and requires much more attention than we have given it so far.

5.7 Conclusion

The results achieved so far are very encouraging, and we are continuing to pursue the original goal.

This paper has mentioned many deferred research tasks and issues. We intend to return to all of them at some time in the future, but at present our highest priority is the investigation of control.

---

[4] One advantage of predetermined paradigm decompositions, when well designed, is that they do facilitate consistency checking. We would like to be able to explain what works and why.

References

[1] Robert Balzer, Thomas E. Cheatham, Jr., and Cordell Green. *Software technology in the 1990's: Using a new paradigm.* IEEE Computer XVI(11):39-45, November 1983.

[2] Bill Curtis, Herb Krasner, and Neil Iscoe. *A field study of the software design process for large systems.* Communications of the ACM XXXI(11):1268-1287, November 1988.

[3] Mike Gordon. *Varieties of theorem provers.* Position paper distributed at the workshop.

[4] Brent Hailpern. *Multiparadigm languages and environments.* IEEE Software III(1):6-9, January 1986. Guest editor's introduction to a special issue.

[5] David Harel. *Statecharts: A visual formalism for complex systems.* Science of Computer Programming VIII:231-274, 1987.

[6] C. A. R. Hoare. *Communicating Sequential Processes.* Prentice-Hall International, 1985.

[7] International Organization for Standardization and International Electrotechnical Commission. *LOTOS—A formal description technique based on the temporal ordering of observational behavior.* Joint Technical Report DIS 8807, 1988.

[8] Michael Jackson. *Description is our business.* In VDM '91: Formal Software Development Methods (Proceedings of the Fourth International Symposium of VDM Europe), pages 1-8. Springer-Verlag, ISBN 3-540-54834-3, 1991.

[9] M. A. Jackson. *Principles of Program Design.* Academic Press, 1975.

[10] Michael Jackson. *Some complexities in computer-based systems and their implications for system development.* In Proceedings of CompEuro '90, pages 344-351. IEEE Computer Society, ISBN 0-8186-2041-2, 1990.

[11] Michael Jackson and Pamela Zave. *Conjunction as composition.* In preparation; draft available from the authors.

[12] Michael Jackson and Pamela Zave. *Domain descriptions.* In preparation; draft available from the authors.

[13] Cliff B. Jones. *Systematic Software Development Using VDM*, pages 231-239. Prentice-Hall International, 1986.

[14] Van E. Kelly and Uwe Nonnenmann. *Inferring formal software specifications from episodic descriptions.* In Proceedings of the Sixth National Conference on Artificial Intelligence, pages 127-132. American Association for Artificial Intelligence, ISBN 0-934613-42-7, 1987.

[15] Leslie Lamport. *A simple approach to specifying concurrent systems.* Communications of the ACM XXXII(1):32-45, January 1989.

[16] Jeannette M. Wing. *A specifier's introduction to formal methods.* IEEE Computer XXIII(9):8-24, September 1990.

[17] Pamela Zave. *A compositional approach to multiparadigm programming.* IEEE Software VI(5):15-25, September 1989.

[18] Pamela Zave. *The operational versus the conventional approach to software development.* Communications of the ACM XXVII(2):104-118, February 1984.

6 Integrating Methods in Practice

Anthony Hall

6.1 Introduction

Praxis is a software engineering company. We develop software-based systems, and we want to do so in a controlled and predictable way, as an engineering discipline. We are therefore very concerned about the methods we use to carry out development. We carry out a wide range of projects and among them are several, large and small, where we have used formal methods. We use formal methods not because we are told to but because we think that they offer us advantages over alternative approaches. It is important to us to understand what those advantages are (if indeed they are real), how they come about, and what their limitations are.

We are currently carrying out a large project developing a distributed real time system where we have used formal methods in the requirements, system specification and design. This provides illustrations of many of the strengths and weaknesses of formal methods.

It is not enough, however, to collect experience; if we are to apply this experience in the future we need to have some explanation of what we have found. I believe that there is a rational basis for our findings. By looking at what we need to do during a development, and by seeing what contribution each method makes, we can see where to use particular methods and, indeed, where we do not have satisfactory methods for what we want to do.

Section 6.2 describes, in very general terms, the problem of developing a large system. Within this framework it describes a general notion of method and some criteria for deciding what constitutes a good method.

Section 6.3 concentrates on those aspects of development normally classified, at least within the formal methods community, as *specification*. It explains, in particular, why specification is an important step in development, the advantages of formal over informal specifications, and also why other methods, sometimes regarded as antithetical to formal methods, can also be effective and are sometimes essential.

Section 6.4 describes some of our experiences in the design stage of the project. Here we find a large number of different kinds of decision are needed. Although there are formal methods for design, they address only a fraction of the total design problem. Formal methods are undoubtedly useful in establishing the correctness of certain kinds of design step. However, there are many other design decisions which we cannot yet formalize, and in any case formal methods do not tell us whether even a correct design step is actually a useful one.

Section 6.5 draws some conclusions from our experience by suggesting those areas which need further research. It contains both positive and negative suggestions: areas where I believe research could, in the short term, be of real help to the practitioner and areas where I believe that current research ideas may, at least in the short term, be going down blind alleys.

6.2 Development and Development Methods

The system development problem is simply stated: starting from some *requirements* we must produce an *implementation*.

To see why this is difficult, we need only look at the characteristics of the start and end points of the process. There are four axes we can use for classification and on each of them requirements and implementation are at opposite ends:

1. precision;

2. level of detail;

3. user-orientation vs computer-orientation;

4. what vs how.

We start with the users' requirements. These are typically:

1. imprecise: users rarely know exactly what they want;

2. high level: they give general descriptions of what is required, with little detail;

3. expressed in terms of concepts from the users' domain, rather than in programming terms;

4. describing *what* is required, not *how* it is to be achieved.

From these we must develop an implementation which is

1. precise: there is no such thing as a fuzzy program;

2. detailed: every single aspect of its behaviour is exactly determined, and this in turn means that, in a large system, it is very big and very complicated;

3. computer-oriented: the implementation is expressed in some programming language; real-world objects have disappeared, and been replaced by strings and database records;

4. algorithmic: the code of the implementation says how a result is to be achieved, but does not say what the result should be.

System development is the transformation between these two representations. There are two points to notice:

1. The changes are almost all for the worse. In particular, the implementation is much harder to understand than the requirements. This is, of course, the reason for the massive reverse engineering industry which is springing up.

2. The transformation is necessarily creative. There can be no effective procedure for generating an implementation from a requirements statement.

In practice, the way we solve the system development problem is to divide and conquer: we produce a number of intermediate representations. For example we may move from requirements to a high level design, then to detailed design and then to code.

A development method, then, is a prescription for:

- what intermediate representations to produce;

- what order to produce them in

- how to produce each representation;

- how to validate each representation.

Of course methods vary widely, not only in what advice they give but also in the degree to which they address the different questions.

A development step consists of replacing one set of representations with another set. Each step replaces:

1. vagueness with precision;

2. summary with detail;

3. user-oriented constructs with machine-oriented constructs;

4. requirements with algorithms.

Now in order to decide what is a good method, we need to understand what is a good design step. There are two characteristics of a good design step: *correctness* and *progress*.

Correctness means that we are not doing anything wrong: that the new representation is consistent with the representation from which it was derived.

Progress means that we are doing something right: we are actually getting nearer the goal of an executable implementation.

Now the correctness of a design step may be (at least semi-) decidable: if the two representations are adequately formal, it may be possible to prove that one is correctly derived from the other. There is, however, no such procedure for deciding on progress - there is no way of telling, until the implementation is complete, whether or not a design step is going down a blind alley.

Having characterised a good design step, we can also characterise a good design representation. A representation is the output of one design step and the input to another. A good representation, therefore, must

- be capable of being checked against a previous representation;

- be usable as the basis for checking subsequent representations.

Note that these two conditions are by no means the same: for example, it is easy to verify that a performance specification is consistent with a requirement, but there is no known way of using this in a subsequent refinement.

6.3   Aspects of Specification

Now that we have some basis for describing and evaluating design representations, we can look more closely at some particular kinds of design representation and some particular methods. One kind of representation which is central to the use of formal methods is the *system specification*. A system specification is characterised by being a statement about *what* the system is to do rather than about how it is to do it. In other words, of the four axes along which we might make design progress, a specification is a representation that does *not* move along the last, requirements to algorithms, axis. Different kinds of specification make different amounts of progress along the other axes: increased precision, addition of detail and increasing machine orientation.

We can ask two questions about specifications. First, when should we write specifications before any other kind of design representation? Second, what kind of specifications should we write? The answers to both these questions are rooted in our characterisation of a good design representation.

The question of what design steps we should take first can be answered by appeal to Barry Boehm's spiral model of development. He suggests that each development step should minimise the risks of subsequent development. Faced with a typical imprecise requirements statement, the main risk that most projects face is that they will build the wrong software: that the user requirements will be in some way misunderstood and the user will reject the final product. A specification should be written, therefore, to clarify in the minds of both developer and client just what it is that is going to be developed.

The question of what is a good specification can be answered by appeal to the same criterion. The risk we are trying to eliminate is that we build the wrong system; therefore we want to be able to tell, as soon as possible, when we are wrong. A good specification, therefore is *refutable* - if it's wrong, you can see that it is wrong. This is, of course, exactly the first criterion which we proposed above for any design representation: that you can tell whether it is correct with respect to its predecessors. The second criterion is just the same idea applied to the next stage of development: given the specification, it must be possible to tell if your implementation is wrong.

A specification, therefore, is like a scientific theory about the system you are trying to build. The characteristic that makes a theory scientific is that it could be refuted, and the more refutable a theory is the stronger it is. So the more ways there are of finding mistakes in your specification, the better a specification it is. Having a good specification early on means you will find mistakes earlier in the project when, as we all know, they are relatively cheap and easy to eliminate.

Formal specifications are, in this sense, good specifications. Formal notations are unique in two ways. First, they allow you to progress along the first axis, increasing precision, without necessarily adding detail and without moving from user-oriented to computer-oriented concepts. Because they are precise, while at the same time still talking the user's language and at a comprehensible level of detail, mistakes are more clearly visible. Second, because you can reason about a formal specification, you can attempt to prove that it has the desired properties. Such proof attempts can, if they fail, reveal errors and also show *why* the specification is inadequate.

However, formal specifications are not the only kind of specifications which are strong because they are refutable. Another very effective technique for finding mistakes in a proposed system is to prototype it. This is because, while users may not know what they want, if you show them something they will quickly tell you whether it is right or not. Prototypes and formal specifications, therefore, work for exactly the same reason: they are both refutable.

This does not mean, however, that prototypes and formal specifications are, as some people seem to believe, mutually exclusive alternatives. On the contrary, in large system development we need to use *every* effective technique we can to reduce risk. We need, therefore, to understand what their domains of application are. I believe that, on the whole, prototyping is the best, possibly the only, way of establishing a good specification of the user interface of a system. Conversely I believe that formal description is the best, possibly the only, way of establishing the underlying state and behaviour of the system.

I can illustrate these ideas by describing the specification of our system and how we arrived at it. The system is a real time information system for a control application. To specify it we used several different representations, each of them aimed at different aspects of the system. The specification consists of three parts.

The first part is the *system context*. This is a definition of what objects exist in the real world that are of interest to the system, how they are related to each other, and what information flows in and out of the system about these objects. The system context is represented using conventional structured methods: entity-relationship diagrams to represent the objects, and data flow diagrams to represent the information flows.

The second part is a definition of the *model* that the system holds about the world, and the operations on that model. This is the part which is specified formally: there is a VDM specification of the data that the system holds, the validity rules for the data, and the effect of each operation. In addition there is a CSP specification of the possible concurrent behaviour of the various operations.

The third part of the definition is the *user interface*. This is defined by example screens and finite state machine descriptions of the possible conversations. This part of the definition was obtained by prototyping and evaluation of the prototypes.

Now a most important fact about these three representations is that they are all essential. None of them is subsumed by the others, and each of them describes different aspects of the system. On the other hand, none of them stands alone. For example, there is a correspondence between each abstract operation described in the VDM and a conversation described in the user interface definition. Furthermore, none of them can be validated on its own. For example, although we can be extremely precise about an operation in the VDM description, we cannot decide whether our set of VDM operations is complete unless we know all the dataflows in and out of the system.

In summary, specification is undoubtedly the area where formal methods are most often applied and best understood. They are a very powerful and necessary technique. However, we must understand that they only address certain aspects of any system, and we must expect to combine them with other methods. Just how different methods fit together is still an open question.

6.4 Aspects of Design

If we have started our development with a formal specification, then we have made progress along the first axis: we have increased the precision of the requirements, to a point where it is agreed just what the system will do. The remaining design steps must move along the other axes: they must

- increase the level of detail;

- translate from user concepts to computer concepts;

- move from what to how.

6.4.1 Detail and Structure

Structuring Methods. As we add detail, the design becomes large and difficult to understand. In order to keep it under control we have to introduce *structure*: we break down the design into parts and describe each part separately. Many design methods are essentially prescriptions for carrying out this decomposition. The paradigm which most methods follow is this:

- Identify components.

- Specify each component.

- Define connections between components.

- Implement each component.

- Integrate the components.

If we are using formal methods, then the second step is done formally. However, formal methods give us no help whatsoever with the first step. It is just this question of how to divide a system into components which is addressed by the various structured methods such as Yourdon, SADT, object-oriented design and so on. Therefore we must look to these methods, as well as formal methods, when we are developing a design.

The division into components is a critical part of the design activity. Not only does it determine the structure and comprehensibility of the design itself, but, more important, it is the basis for the division of work on the implementation. Each team is given a set of components to implement, and their work is defined by the definition of the components.

Problems with Structure. Unfortunately, there are serious problems in carrying out this division.

First, there is no single criterion which can be used to divide a system up. For example, if we follow the object-oriented approach we identify objects - clusters of state and functionality - as our components. If we look at the design from a user interface point of view we identify user tasks or transactions as the components. If we look at the concurrency inherent in the system we identify processes as the components.

Second, if we take any two ways of dividing the system up, we find they do not fit together. A single object may be used in several transactions, and a single transaction may involve several objects and several processes. So we need to consider all the different structures, perhaps using a different method for each, and somehow divide up the work so that at the end of it all everything fits together. Since there is no one natural hierarchy, it is difficult to divide up the work of implementation.

Third, none of the design methods satisfies our criteria for a good design step. Given a specification, there is no way of verifying that a particular collection of components is a correct design for that specification (although it is, fortunately, possible to verify that a subsequent step is correct with respect to the components). Nor is there any way of knowing whether progress is being made - whether the components defined are, themselves, implementable with less effort than the original system. Perhaps the most common design mistake is to propose a collection of components which is either

- unimplementable - the lower level components cannot be built

or

- useless - the system cannot be constructed out of the lower level components.

Such a mistake, early on in a project, can be very hard to find and very expensive.

An Example. In our system we had four dimensions of structure to be reconciled:

1. user tasks;

2. data types;

3. processes;

4. machines.

Each of these gave rise to a different structure imposed on the software. The user interface code was structured according to the dialogues that the user carried out. The code to manage the major data types was structured according to object-oriented ideas, encapsulating each type of data and the functions relating to it into a single module. Individual processes were identified according to the intrinsic concurrency in the application and also the need for concurrent processing to improve performance. Data, operations and processes were allocated to machines according to performance and availability criteria.

None of these decisions was in any way formal, and it was often difficult to know whether a particular decision was correct or not. Indeed, many of the decisions were made on the basis of informally-specified requirements for non-functional aspects of the system: performance, reliability and so on. Having identified the components, however, we did use formal methods as far as possible to specify them. The data types were defined in VDM, just as the specification data types were. Processes were defined either using finite state machines or, in the case of more complicated processes, using a process algebra, CCS.

6.4.2 Reification and Algorithm Design

The second kind of design step is to move from a user-oriented concept (such as a control valve) to a computer-oriented concept (such as a database record). This aspect of design does have a well-developed formal theory: it is called reification. We can propose a representation of a particular specification construct and we can indeed verify that the representation is correct with respect to the specification. We used this idea on our system to develop the modules which represented the main data types.

Similarly, given a sufficiently detailed specification of a small procedure, it is possible to carry out a formal development of the imperative code to carry out the procedure.

In practice, we found limitations on the how formal we could be in carrying out these steps.

First, there is only a well developed theory for sequential programs. Reification of processes is not yet well understood.

Second, we were implementing on top of proprietary operating systems so that the semantics of our implementation were not formally defined. This means that we could not prove that, for example, a file was an adequate representation of some construct because we did not have a complete semantics for the file operations.

Third, there is an enormous volume of relatively trivial maths involved in carrying out and proving a formal statement of the reification relation. This is exacerbated by the problems of structuring: reification is not compositional if the composition involves shared state.

This last is essentially an economic question: it is possible, but very expensive, to use formality. The question is how to decide whether the benefits outweigh the costs. The answer lies, once again, in thinking of the design as a theory about the system. The question is, "how can I tell if this theory is wrong?" Now most reifications are straightforward and it is quite easy to see why they are correct, and conversely it is likely that one would quickly find any error. If that is the case, the extra benefit from stating the reification relation formally is small - it is reducing an already small probability of error. The main benefit in these cases comes from application of the *concepts* of reification, such as adequacy, rather than from a complete formal treatment. If, however, the reification proposed is subtle or it is not obvious why it is right, then one would get a large benefit from stating and attempting to prove the formal relation.

This is one problem which may be amenable to attack through better tools. The problem is not subtlety but simply volume, and in principle good tools for handling the bulk of the mathematics could greatly reduce the cost of formalism. Even with tool assistance, however, good compositional notations will be necessary so that the problem can be divided into manageable parts.

6.4.3 Summary

Moving from a specification to a design is by no means entirely a formal process. The design of a large system has many aspects, most of them not well enough understood to be formalised. There is no way of producing a design that can cover all the aspects, and a combination of different approaches must be used. Some of these approaches can contain formal elements, but there are many necessary design steps which cannot be justified formally. One major challenge is to find ways of verifying design steps. Another is to understand how the different aspects of design can be reconciled - for example to find ways of combining concurrent and sequential formalisms.

Where a design method with a formal basis is applicable, it should be used in preference to an informal method because it will give greater probability that errors will be found early on. However, the cost of being completely formal increases as the design becomes larger and more detailed. It is important to understand the benefits of formalism so that you can decide when it is worth applying and when the cost would not be justified.

6.5 Implications for Research and Development

I do not think that long term research should be driven by the needs of practitioners, so I do not propose to say anything about such research. However, there is a lot of research and development aimed at producing useful results in the short term, and on the basis of our experience we can identify areas where research is needed and where we would be in a position to use the results directly in our work.

There are also some questions being pursued which are not going to yield useful results in the short term, and which are therefore not likely to be of practical interest to industry at the moment.

I think there are three areas where progress can be made, but in each case we must be careful to ask the right question. These areas are the comparison of methods, relationship between methods and the integration of methods.

The comparison of methods sometimes takes the form of the search for a 'best' method. It is clear that there is no such method. The question "is VDM better than Yourdon?" is not going to be answered. What *is* needed is to understand the complementary roles of different methods: what each one is good for and what aspects it does not cover.

The second topic is the relationship between methods. Since methods address different aspects of design, it is not useful to think of translating between methods. For example it is pointless to ask how we might have translated our dataflow diagrams into VDM, because dataflow diagrams and VDM specifications are about quite different aspects of the problem. On the other hand, we *do* need to understand what the relationship between a dataflow diagram and a VDM specification actually is. They clearly have things in common and it is important to understand what they are. An important question, for example, is how can we tell whether a DFD and a piece of VDM are compatible. Research is needed both to understand how to use formal methods with structured methods and also to know how to combine different formalisms, such as model-based operation specifications with process algebras.

Third, when we know what methods we want to use and how to combine them, we need tool support which not only covers all our chosen methods but also makes the right connections between them. Again, this is not going to be a single, all encompassing, wide-spectrum language, just because no such language will cover all aspects of the problem. The tools need to be adapted to the different notations and methods but at the same time to recognise the points of contact between the methods and to integrate the common elements.

7 Formal methods and product documentation

John Wordsworth

7.1 Introduction

The proper use of formal methods in software development strongly influences the nature and relationships between the various kinds of documentation produced. Rather than being a mass of heterogeneous information produced mainly for immediate use and then discarded, product documentation should be a coherent collection of information of lasting value in the continuing life of the product. In this paper we look first at the nature of product documentation in a mature product developed with the aid of formal methods. Next we consider how different people in a software development organisation interact with product documentation in their work. Finally we examine the problems that face managers of existing products who wish to construct product documentation in the mould described here.

7.2 The fully formalised software product

In this section we look at the product documentation that would accompany a software product developed *ab initio* using a formal development method such as that described in [5] or VDM [3]. The use of a formal method and the recording that accompanies it, encourages, or perhaps even forces, the adoption of a certain style of documentation. The different documents produced during development have specific roles to play, and these are now summarised.

7.3 The elements of product documentation

Product documentation is of five kinds.

- requirements

- specifications

- data designs

- algorithm designs

- source modules

Requirements are the requests that led to the product being developed. They are usually expressed in an imprecise language such as English. The inclusion of tables and diagrams can add something to the precision of requirements expressed in a natural language. In a mature product, there are many sources of requirements. Existing users of the product will make suggestions about changes to the product, as well as more imperative requests for changes to remove errors, which we discuss below under service. Requirements might arise from the imposition of standards by an external body. The trends of the market place that the product is intended to serve will also generate requirements for it. Where a product is part of a coordinated solution offered by a supplier, developments in the supplier's other offerings, hardware and software, might also produce requirements. It is important that requirements are seen as part of product documentation, even though they are only imprecise. We shall not go into detail about a requirements management process, but simply say that requirements should be kept in a library, and this library is an important part of product documentation. As requirements are received they will be added to the library, unless they are duplicates of existing requirements, or restatements of requirements that the product already meets. The library search facility will probably include the searching of unformatted text for various words and phrases, as well as more structured indexes to its contents. There will be a library of specifications of all the external interfaces that the product provides. A specification provides a model of the data behind an interface, and an explanation of the permitted operations in terms of the model. Specifications should be recorded in a formal language such as Z, and should contain proofs of the consistency of the model, and of the preconditions of the operations. The contents of the specification library will be produced at an early stage of the development process. Specifications are constructed to make precise the functional aspects of the requirements, and can be subjected to various formal and informal analyses, such as consistency (formal) and usability (informal). There will be an explicit relation between requirements and the specifications that satisfy them. This is a most important aspect of product documentation that is too easily neglected. It should be possible from a requirement in the requirement library to discover which aspects of the external interface are alleged to meet the requirement. Conversely it should be possible from a specification to discover which requirements it is meeting. This is of particular value if it is intended to withdraw or change a piece of function, so that the effect on customers can be forecast. In the development process the validation of a specification means establishing a high degree of confidence that the specification does in fact meet the requirements that are alleged to have given rise to it. In the specification library, and other libraries to be described later, there might be problems of version control. In a long-lasting product there might be several releases of the product currently in service, and each should have its own version of the specification library. A data design explains how the data model of the specification is to be represented by components that are in some sense more concrete than the specification. It also explains how the operations in the specification are to be represented in terms of the models appropriate to the components. If a formal development process is being carried out, then for each specification there will be one or more data designs. The components of a data design might be:

- externally provided services (e.g. from an operating system)

- internally provided services (e.g. storage management, resource managers, etc.), and

- primitives of the programming environment (e.g. strings, binary numbers, and arrays).

A data design refers to specifications of the components that it proposes to use in its representation. For those components that are internally provided interfaces, the specifications are in turn related to their own data designs. For the components that are externally provided interfaces the data design should refer to precise specification documents for the externally provided interfaces. If these are not available, then the data design should document the assumptions that are being made about the function of the externally provided interface. Data designs could be recorded using a formal language such as Z, and the proofs of the consistency of the concrete state, and of the correctness of the operations must be present in the data design. Data designs should be kept in a library. The relation between specifications and data designs should be explicitly maintained in product documentation. The multiplicity of data designs for a given specification arises from several causes:

- The product is offered for several different hardware bases, each requiring a different data design.

- Different releases of the same product might need different data designs.

- Choice of data design might be a customer option (e.g. a high performance option, or a multiple user option).

The choice of data design for a given specification depends on non-functional requirements such as performance, auditability, etc. Each operation in a data design has an algorithm design associated with it. The algorithm design shows how the operations provided by the components can be marshalled to realise the operations promised by the specification. Algorithm designs could be made formal by using Z and the guarded command language. to record them. Examples of how to do this are given in [5]. The proofs of the correctness of the refinement steps in the algorithm is part of the algorithm design. The relation between algorithm designs and the data designs from which they arise should be an explicit part of the product documentation. For a given operation in a data design there might be several algorithm designs. The multiplicity of algorithm designs for a given operation arises from similar causes to the multiplicity of data designs for a given specification. Associated with each algorithm design is the source module that implements it. A source module might contain the implementation of several algorithms. Source modules are already formal since they are recorded in a programming language. Source modules are the input to the process that builds the product for distribution to customers. Source modules should be kept in a library, and the relation between source modules and the corresponding algorithm designs should be explicitly maintained in product documentation. There is an approach to development that dispenses with the source module, and regards the algorithm design as the end product of software engineering. In this approach the algorithm design process continues until the design is expressed in sufficient detail for its parts to be translated mechanically into the source of the target programming language. When the product is to be built, a tool scans the algorithm designs, builds the source modules from the fragments of design, and presents them to a compiler. The source modules have only a fleeting existence; they are temporary work files in a single process that transforms algorithm designs into products. This approach reduces the emphasis on a programmer's language skills, and emphasises skills in design. The following diagram illustrates the relations between the formal elements of product documentation.

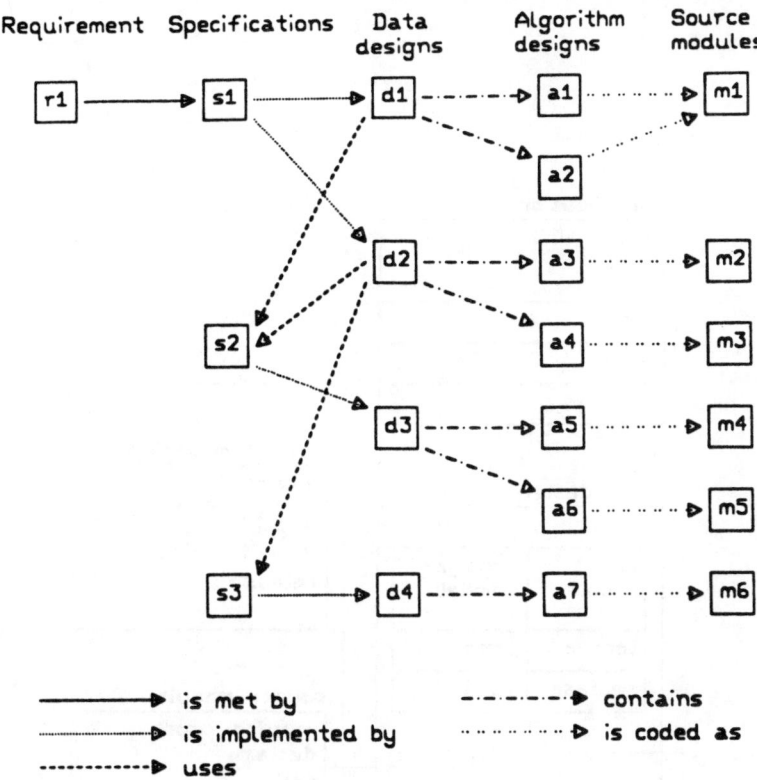

In this diagram the specification of an external interface is represented as consisting of a model and a number of operations, "in", "read", and so on. One of the data designs corresponding to this specification is represented in a similar fashion. However the model is expressed in terms of the components "cpt_1", "cpt_2" that will be used in its construction, and the operations have names that are different from those of the specification, though they implement corresponding functions. Thus "inq_1" is intended to implement the function specified as "inq", and "read_1" is intended to implement the function specified as "read". One of the concrete operations, namely "inq_1", is shown as having an algorithm design in which the operations of the components of the data design are used to implement the concrete operation. The algorithm design has associated with it a source module written in an appropriate language. Each of the components "cpt_1" and "cpt_2" used in the data design will have its own data designs, algorithm designs, and source modules separately recorded. Another, somewhat more remote, view of product documentation is given in the next diagram.

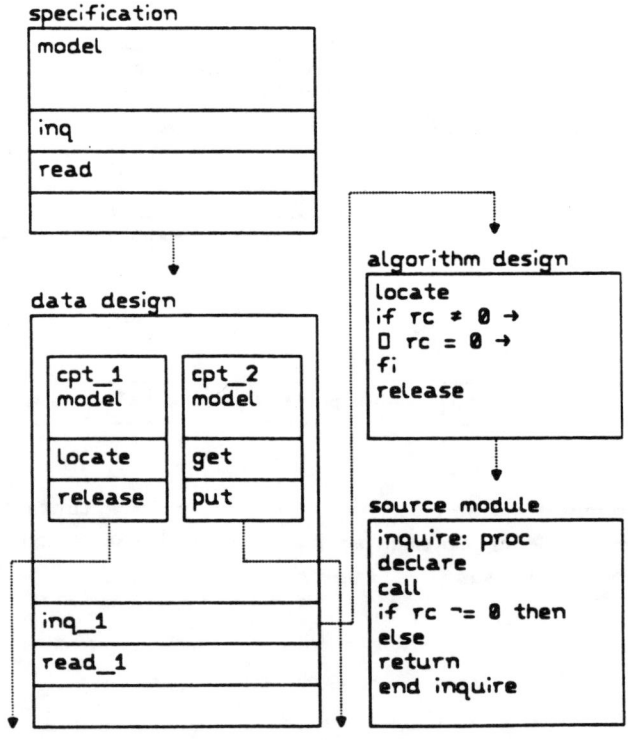

   This diagram shows how a single requirement "r1" might be met by a single specification "s1". Two data designs "d1" and "d2" are provided for this specification. The first data design "d1" implements the specification "s1" by using the facilities of component "s2". There are two operations to be implemented, and each gives rise to an algorithm design. The two algorithm designs "a1" and "a2" are packaged in a single source module "m1". The second data design "d2" implements the specification "s1" by using the facilities of two components "s2" and "s3". It also has two algorithm designs "a3" and "a4", but these are separately packaged in source modules "m2" and "m3". The specification "s1" is the only external interface. Specifications "s2" and "s3" are specifications of components of the product that are not available externally. Each has its own data design, and the data designs have their own algorithm designs and source modules. The data designs "d3" and "d4" use only the facilities of the programming environment to implement their specifications. No externally provided services are used. Different styles of arrows are used to represent the various relations between the elements of product documentation, and names for these relations are suggested.

7.4   Product documentation and the product range

Often a product development organisation has to manage the development of a range of products for different environments. The intention behind offering this range of products is to provide the same external function in different environments. The data designs, algorithm designs, and source modules of these products might differ, but the specifications are intended to be the same. The existence of specifications in product documentation will allow the development team to control any deviations in specification by highlighting the fact that such deviations exist, or are proposed, and will encourage convergence of products that have been allowed to diverge. The existence of data designs and algorithm designs will encourage research by one product into another, and increase opportunities for sharing designs and code.

7.5   Product documentation and product development

The requirements, specifications, data designs, algorithm designs and source modules would be a prime source of information for everyone concerned with the development of the product. The following list summarises the activities in the development process that interact with product documentation.

- Requirements Analysis-relating requirements to existing interfaces and identifying the need for new or changed interfaces

- Planning-assessing costs and benefits of meeting requirements, batching up new function into complete releases, aligning it with operating system releases, etc.

- Architecture-assessing the effect of new function on existing data models and operations, and making or revising specifications

- Design and implementation-making or revising data designs, algorithm designs and source modules

- Function Verification-verifying the correct implementation of the specifications and other changes to the interfaces by writing test cases based on the specifications

- Testing-testing the function of delivered code, and reporting errors

- Performance-analysing the performance of the delivered code against a performance plan, reporting discrepancies, and where possible identifying areas of concern in the design

- Technical Writing-creating user manuals from specifications, or program logic manuals from data designs and algorithm designs

- Service-fixing day-to-day problems in the implementation by studying source modules and their design context

(If the Cleanroom [4] method of development is followed, the Function Verification and Testing activities are replaced by a usage-based statistical test called System Certification.) The use of product documentation in development tasks is considered further in 7.7.

7.6 Product documentation and customers

Some parts of product documentation might affect the relation between the product and its customers. In the past, customers have explored the nature of a product in five ways:

- by going on courses offered by the suppliers or others

- by reading the manuals provided by the suppliers

- by reading the source code (when source code is available)

- by doing their own experiments with the product

- by comparing notes with other users

The existence of product documentation means that precise information about the product could be made available to customers, either in the form of improved manuals, or as specifications. By giving customers a precise description of the function supported, a supplier makes a commitment to provide that function, and no other. In endorsing a specification, a customer makes a commitment to expect that function, and no other. Exploration of the delivered product might well reveal other behaviour of the product that is no part of the supplier's intention, but the customer can refer to the specification to see whether this behaviour is specified, and can therefore judge the supplier's obligation to maintain it. What is not in the specification is not guaranteed. This approach is not possible if specifications are imprecise. Imprecise specifications encourage experiment, and encourage users of a software product to rely on the results of experiments rather than on any stated intention of the function of external interfaces. Customers should respond well to precise specifications of interfaces.

7.7 Developing from a fully formalised base

Changes to a products might arise from several causes:

- new or changed customer requirements

- mistakes in the implementation detected before release

- mistakes in the implementation detected after release

- second thoughts about the best way to implement functions

- the need to migrate software from one environment to another

7.8 New requirements

When a requirement arises it will be compared with existing requirements to decide if it is a new requirement not already met. This comparison might be done by considering informally which external interfaces are most likely to be concerned in meeting the requirement. From the specifications of these interfaces it will be possible to trace the existing requirements that are met by the specifications. If it is a new requirement it will be assessed on business grounds to see if it is worth meeting. This assessment will include some work on the likely cost of meeting it, which will depend on the changes to specifications, data designs and algorithm designs entailed. It will also depend on a technical assessment of the kind of release environment into which such a change will fit requisite product changes, operating system environment, requirements of other products, likely pattern of service pressure, and early support programs, etc. The details of the specification changes will be worked out, and this will involve some more detailed planning of the likely changes to data designs and algorithm designs. A new requirement might be met wholly or in part by changes to existing specifications, or it might require the creation of new specifications. The specifications will then be used in the development of the data designs, algorithm designs, and source modules. Of course the software engineer in charge of this development will be concerned to reuse as much as possible of the existing data designs, algorithm designs, and source modules. The explicit relationship between the specifications and data designs will ensure that the engineer does not miss other data designs implementing the same specification. The non-functional requirements of the specification such as performance, operating system environment, hardware, etc., will all have an influence on the decisions made in data and algorithm design. The product documentation relating to this requirement (specifications, data designs, algorithm designs, and source modules) is subject to function verification to see if any discrepancies have crept in. The functions of the interfaces are tested against their specifications to see if they agree. It is important to check that old functions still present in the specification have not been lost by redevelopment. The discipline of the development process should prevent this kind of regression from occurring. Testing has regard to the boundary values of the data and algorithm designs as well as to the boundary values of the specification. The assessment of the performance of the product might point the way to changes needed in the data and algorithm designs, or occasionally in the specification. Certain design choices might prove in the event to have been inept, or certain functions might prove to be too expensive in terms of resource consumption. The role of technical writers should not be confined to creating publications for the product's users; they should also be concerned with the creation of product documentation itself. The creation of user documentation by technical writers is much helped by the existence of product documentation. The kinds of models used in the specifications should be made explicit in the user documentation, and the operations should be described in the same way as in the specification. The exposure of readers to formal text should certainly be considered.

7.9   Errors detected before release

Errors will be detected by Function Verification and System Test. In a formally developed product the aim is defect prevention rather that defect detection, but there might be errors to be detected at this stage. These errors arise from mistakes in understanding the requirements (specification errors), mistakes in doing the proofs of correctness (if these are done by hand), or errors of judgment in deciding which refinements to prove correct, and which to leave unproved. Causal analysis is the process of determining the origin of each error, whether in specification, data design, algorithm design or source module, and the results of this analysis are used to improve the quality of the development process. Errors must be fixed from their origin downwards. It is essential to maintain consistency in the relations between the parts of the product documentation, since so many parts of the development process rely on it.

7.10   Errors detected after release

After the function has been shipped, errors might be discovered by customers. These errors arise from the same causes as those discussed in the previous section. Causal analysis determines the origin of the error, and a temporary fix or bypass is produced. The origin of the error, and the fix or bypass, must be reported to the owner of the material in which the error originated. Maintaining consistency of the product documentation is no less important after release than it was before release. Although a particular release and its specification have a limited life, changes to the product documentation for one release need to be migrated into the product documentation for the next. If a fix requires some change to a part of the product documentation, the change should be made in consultation with the owners of any parts of the product documentation affected. In particular if the fixes result from specification errors, the external interface of the product will change, and users of it need to be informed. If customers are not informed of changes to the specification they might rely on experiments and hearsay to determine the function of the product.

7.11   How to get there

Many existing products do not have fully-formalised product documentation of the kind described here. The construction of product documentation for an existing product presupposes that

- **either** there is a rational explanation for what exists, and it is economic to seek it out and write it down,

- **or** it is economic to redevelop the parts of the current implementation for which a rational explanation cannot be found.

The assumption that there is a rational explanation for what exists is one made every day by any one who has to understand any part of the product. In this class we must include all the users of the product, and all the members of the development organisation who are trying to adapt the product to meet new requirements. For lack of product documentation much exploration and experiment has to be undertaken. The practice of writing down what has been discovered is not usual. The published manuals attempt to provide a rational explanation of what exists, but they generally lack the precision of formal text. The benefits to be had from throwing away some of what exists and replacing it by something that has a rational explanation was one of the motivations of the CICS restructure project [1]. Existing interfaces should be documented by writing specifications, and their current implementations should be documented by writing data designs and algorithm designs. The information generated and discovered by this process should become product documentation, and be stored in the appropriate libraries as outlined above. The creation of specifications of existing interfaces is a time-consuming business, and throws an interesting sidelight on the nature of software and of human interactions with it. It is discussed in detail elsewhere [2]. The difficulty of disentangling the intended function of the product from its actual, supposed, and documented behaviour suggests that it might be better to respecify a new product than to attempt to convert an old one. Continuing development of an existing product has a deleterious effect on efforts to construct product documentation for it.

7.12 Summary

Product documentation, a coherent and precise account of the current state of a piece of software, is a valuable asset in a software development business. Its existence affects the method of working of everyone who is responsible for developing and servicing it. In this paper we have explored its effect on the people who use it, and looked at some of the problems of creating it for an existing product. We have illustrated a development method that, as better tools are developed, could allow developers and customers to derive significant benefit from formally based product documentation.

7.13 Acknowledgments

I am grateful to various colleagues at IBM's Hursley laboratory for discussions leading to the ideas set out in this paper, and I should like to thank Peter Lupton and Lionel Walker for suggesting improvements to the text.

References

[1] B. P. Collins, J. E. Nicholls, I. H. Sørensen Introducing formal methods: The CICS experience with Z

IBM Hursley Technical Report TR12.260

[2] I. S. C. Houston and S. King *Experiences and results from the use of Z in IBM* Proceedings of VDM '91, Springer-Verlag, Lecture Notes in Computer Science 551, August 1991

[3] C. B. Jones *Systematic Software Development Using VDM* Prentice-Hall International, 1989

[4] H. D. Mills, M. Dyer and R. C. Linger *Cleanroom Software Engineering* IEEE Software, September 1987, pp19-25

[5] J. B. Wordsworth *Software Development with Z* to be published by Addison-Wesley, 1992

8   Software Quality : A Modelling and Measurement View

Victor R. Basili

8.1   Software Quality Needs

In order to deal with the concept of software quality, we need deal with such issues as quality definition, process selection, evaluation, and organization. In order to develop a "quality" object, we need to define the various qualities and quality goals operationally relative to the project and the organization. It is clear that there are a large number of qualities or characteristics that a product may possess and different projects will prioritize these qualities differently. Once we have defined the kind and level of qualities we want, we need to find criteria for selecting the appropriate methods and tools and tailoring them to the needs of the project and the organization in order to achieve these quality characteristics. This implies we may be interested in the quality of the processes as well as the quality of the product. We need to evaluate the quality of the process and product relative to specific project and organizational goals while the product is being developed, as well as when it is complete. This is to assure we know we are on the right track and can make the appropriate adjustments to the process to achieve our goals. Lastly, we must create a support organization to oversee the quality goals from planning through execution, by evaluation, feedback and improvement.

The software industry has been slow in recognizing these needs for software quality. Part of the problem has been a lack of understanding of the nature of the artifact and its desirable qualities and the difficulty of achieving the quality characteristics desired. In many organizations, the definition of quality is simply a small number of errors reported from the customer. This is a rather narrow and limited definition of quality since quality means more than error reports. Most companies have a software development methodology standards document (which few projects follow). For them, process selection means using this single model of process definition, independent of the project characteristics and quality needs. That is, using the same processes for a well understood project development and a first time product. Quality evaluation often means counting the number of customer trouble reports generated. This is a rather passive view of quality, not an active one. It does not provide a mechanism for learning how to produce software with the desired characteristics and does not provide an evaluation of the methods or tools needed to provide those characteristics. Lastly, for many companies the quality organization is the group that provides management with the trouble report data. This is not a quality oriented organization, quality is the result not the driver.

What kinds of quality approaches exist? We can have quality control and quality assurance. Quality control is the act of directing, influencing, verifying, and correcting for ensuring the conformance of a specific product to a design or specification. Quality assurance is the act of leading, teaching and auditing the process by which the product is produced to provide confidence in the conformance of a specific product to a design or specification. Quality control is project oriented, quality assurance is process oriented.

Quality control involves such activities as evaluating products and providing feedback to the project. A quality control organization is typically part of the project and interactive with it. It requires people with a knowledge of the processes and the product requirements, and an understanding of the solutions.

On the other hand quality assurance usually involves the following activities:

- Definition and evaluation of processes

- Collection of data from quality control

- Feedback to the projects it is assuring, as well as to the organization and itself (for the purpose of learning).

It involves an independent chain of command from the project. People requirements involve a high level of knowledge with respect to technology and management and a deep understanding of the process and the product.

What control do quality control and quality assurance have? How often is a project stopped from moving on to the next phase? How often is a design rejected because it doesn't pass standards? The answer is not too often. This is because we do not have the information necessary to make the proper decision, i.e., we do not have models and baselines (measurement data) to provide the real understanding of the situation to exercise control.

8.2   Modelling Software Experiences

In the software business, we need to be able to characterize and understand the the elements of the business. We need to be able to differentiate project environments and understand the current software process and product. We need to provide baselines for future assessment. We need to build descriptive models of the business.

We need to be able to evaluate and assess our processes and products. That is we need to assess the achievement of quality goals and the impact of technology on products. We need to understand where technology needs to be improved, tailored. We need to be able to compare models.

We need to be able to predict and estimate the relationships between and among processes and products. We need to find patterns in the data, product and processes that allow us to build predictive models.

We need to motivate, manage, and control the software development process by providing quantitative motivation and guidelines that support what it is we are trying to accomplish. We need to build and use prescriptive models.

These issues argue for the need to define models to help us understand what we are doing, provide a basis for defining goals, and provide a basis for measurement. We need models of the people, e.g., customer, manager, developer, the processes, and the products. We need to study the interactions of these models.

These models should be as formal as we can make them, based upon the maturity of our understanding of the software process and products. Once we have models we can ask the following questions: Is the model correct in principle? Does the model actually describe what we are doing? How can we improve the model based on theory, practice and analysis? How do we feed back what we have learned to improve the model or our adherence to it? We want to build descriptive models to explain what is happening. We want to define prescriptive models to motivate improvement.

What kinds of things can we package into models? At NASA/GSFC, in the Software Engineering Laboratory, we have built resource models and baselines, e.g., local cost models, resource allocation models, change and defect baselines and models, e.g., defect prediction models, types of defects expected for the application, project models and baselines, e.g., actual vs. expected product size, library access, over time, process definitions and models, e.g., process models for Cleanroom, Ada waterfall model, method and technique evaluations, e.g., best method for finding interface faults, products and product parts, e.g., Ada generics for simulation of satellite orbits, quality models, e.g., reliability models, defect slippage models, ease of change models, and lessons learned, e.g., risks associated with an Ada development.

Many of the models we use are based on measurement. Measurement takes on different forms. We perform objective and subjective measurement. Objective measures are absolute measures taken on the product or process, usually involving interval or ratio data. Examples include time for development, number of lines of code, work productivity, and number of errors or changes. Subjective measures represent an estimate of extent or degree in the application of some technique or a classification or qualification of problem or experience. There is no exact measurement and the measures are usually done on a relative scale (nominal or ordinal data). Examples include the quality of use of a methodology or the experience of the programmers in the application.

We use a mechanism called the Goal/Question/Metric, GQM [BaWe84], paradigm to build particular models based upon goals. A GQM is a mechanism for defining and interpreting operational, measurable software goals. It combines models of an object of study, e.g., a process, product, or any other experience model, and one or more focuses, aimed at viewing the object of study for characteristics that can be analyze from a point of view, e.g., the perspective of the person needing the information. This orients the purpose of the analysis, namely, to characterize, evaluate, predict, motivate, improve, specifying the type of analysis necessary to generate a GQM model relative to a particular environment. In this context, goals may be defined for any object, for a variety of reasons, with respect to various models of quality, from various points of view, relative to a particular environment.

We must build models of our various products and processes. By way of example, consider the need to build an operational model of our education and training in a particular process. We articulate the process as the following set of activities:

1. Provide the individual with training manuals they must read.

2. Provide a course, educating the individual in the process.

3. Provide training by applying the process to a toy problem.

4. Assign the individual to a project using the process, mentored by an experienced method user.

5. After this the individual is considered fully trained in the process.

We convert into an operational model by associating a set of interval values with the various steps of the process. In this case, since the model is clear, each of the steps represents a further passage along the interval scale. Thus a value of "0" implies no training,

1. implies the individual has read the manuals

2. implies the individual has been through a training course

3. implies the individual has had experience in a laboratory environment

4. implies the process had been used on a project before, under tutelage

5. implies the process has been used on several projects.

Even though we call this a subjective rating, if the education and training process model is valid, then our model and the metrics associated with it are valid. Using the GQM, we generate a question that gathers the information for the model.:

Characterize the process experience of the team (subjective rating per person) : (again, "0" corresponds to no training)

1. have read the manuals

2. have had a training course

3. have had experience in a laboratory environment

4. have used on a project before

5. have used on several projects before

The data from the question can then be interpreted in a variety of ways, e.g., if there are ten team members, we might require that a minimum requirement is that all team members have at least a three and the team leader has a five, etc. This evaluation process will become more effective with experience over time.

Factors or characteristics that affect software development and which need to be modelled include people factors such as number of people, level of expertise, group organization, problem experience, process experience, problem factors such as, application domain, newness to state of the art, susceptibility to change, problem constraints, process factors such as, life cycle model, methods, techniques, tools, programming language, other notations, product factors such as, deliverables, system size, required qualities, e.g., reliability, portability, and resource factors such as, target and development machines, calendar time, budget, existing software.

These models help us to classify the current project with respect to a variety of characteristics, to find the class of projects with similar characteristics and goals to the project being developed, and to distinguish the relevant project environment for the current project. They provide a context for goal definition, reusable experience/objects, process selection, evaluation, comparison, and prediction.

The choosing and tailoring of an appropriate generic process model, integrated set of methods and techniques is done in the context of the environment, project characteristics, and goals established for the products and other processes [BaRo87]. Thus, the model for expressing process needs to provide a flexible process definition appropriate, information for process selection, and support for process integration and configuration, via tailorable definitions and characterizations for life cycle models, methods and techniques.

Goals need to be defined for the various processes. They help in the choice of a life cycle model, methods, and techniques. To show why this is necessary, consider the testing phase of a project development. What are the goals of the test activity? Is it to assess quality or to find failures? The selection of activities depends upon the answer. If it is to assess quality, then tests should be based upon user operational scenarios, a statistically based testing technique, and reliability modelling for assessment. If the goal is to find failures, then we might test a function at a time, general to specific. Reliability models would not be appropriate.

For example, it needs to permit advice and selection criteria based upon the problem characteristics and goals, containing such rules as:

- If the problem and solution are well understood, choose the waterfall process model

- If a high number of faults of omission expected, emphasize traceability reading approach, embedded in design inspections

- When embedding traceability reading in design inspections, make sure traceability matrix exists.

8.3 Model Evolution

The research in software engineering typically involves the building of technologies, methods, tools, etc. Unlike other disciplines, there has been very little research in the development of models of the various components of the discipline. The modelling research that does exist has centered on the software product, specifically mathematical models of the program function.

We have not emphasized models for other components, e.g., processes, resources, defects, etc. What is needed is a top down experimental, evolutionary framework in which research can be focused, logically and physically integrated to produce models of the discipline, that can be evaluated and tailored to the application environment.

In the SEL, we use an organizational framework called the Quality Improvement Paradigm [Ba85a]. It consists of the following steps:

- Characterize the current project and its environment with respect to models and metrics.

- Set the quantifiable goals for successful project performance and improvement.

- Choose the appropriate process model and supporting methods and tools for this project.

- Execute the processes, construct the products, collect and validate the prescribed data, and analyze it to provide real-time feedback for corrective action.

- Analyze the data to evaluate the current practices, determine problems, record findings, and make recommendations for future project improvements.

- Package the experience in the form of updated and refined models and other forms of structured knowledge gained from this and prior projects and save it in an experience base to be reused on future projects.

We have used this evolutionary paradigm to define, refine and evolve models in the SEL. For example, we have used it to help develop and formalize the definition of process models as follows:

- Use the project characteristics and goals to find the most appropriate process models for the current project

- Develop, modify or refine the process based upon the lessons learned from previous applications of the model

- Set goals to monitor those new or high risk areas

- Execute, collect and analyze data, making changes to the process in real time

- Based upon goals and analysis, write lessons learned and modify the process for future use

One such model has been the Cleanroom Process Model [Dy82], proposed by Harlan Mills. The key components of the process model are a mathematically based design methodology, the functional specification for programs, state machine specification for modules, reading by stepwise abstraction, correctness demonstrations when needed, and top-down development. Implementation is done without any on-line testing by developer. Testing is performed by an independent test group using statistical testing techniques, based on anticipated operational use. Testing takes on a quality assurance orientation, rather than a failure finding orientation.

Before applying the Cleanroom in the SEL environment, we ran a controlled experiment at the University of Maryland [SeBaBa87]. The goal of the study was to analyze the Cleanroom process to evaluate it with respect to the effects on the process, product and developers relative to differences from a non-Cleanroom process. The project was an electronic message system (1500 LOC) and there were 15 three-person teams (10 used Cleanroom). Each team was allowed 3 to 5 test submissions. We collected data on the developers background and attitudes, all on-line activity, and the test results.

The results of the Cleanroom project were quite strikingly in favor of Cleanroom. The Cleanroom developers felt they more effectively applied off-line review techniques, while others focused on functional testing. The Cleanroom developers spent less time on-line and used fewer computer resources. The Cleanroom developers tended to make all their scheduled deliveries. The product developed by the Cleanroom teams had less dense complexity, a higher percentage of assignment statements, more global data, and more comments. The Cleanroom products more completely met requirements and had a higher percentage of test cases succeed.

Based upon the success of the controlled experiment, we decided to experiment with the Cleanroom process model in the SEL. Following the Quality Improvement Paradigm, we

- Used the project characteristics and goals to find the most appropriate process models for the current project, i.e. we picked the process based upon the project needs and process strengths, e.g., Cleanroom for better lowering defect rate.

- Modified and refined the process based upon the lessons learned from previous applications of the model. Existing process model descriptions available for use included the standard SEL model, IBM/FSD Cleanroom Model, experimental UoM Cleanroom model. Lessons learned associated with the IBM/Cleanroom model included: basic process model, methods and techniques, and knowledge that the process was very effective in given environment. Lessons learned from the UoM/Cleanroom model included the facts that no testing enforces better reading, the process was quite effective for small projects, formal methods were hard to apply, require skill, and we might have insufficient data to measure reliability. We defined the SEL/Cleanroom process model using the best practices from our experience base models, e.g., informal state machine and functions, training consistent with UoM course on process model, methods, and techniques, emphasize reading by two reviewers, allow back-out options for unit testing certain modules, etc. When no new information was available, used the standard SEL activities.

- Set goals to monitor those new or high risk areas. Here the goals were to characterize and evaluate in general, and with respect to changing requirements.

- Execute, collect and analyze data, making changes to the process in real time. We monitored the project carefully and made changes to the process model in real time.

- Based upon goals and analysis, write lessons learned and modify the process for future use. We wrote lessons learned for incorporation into next version, redefining process for the next execution of the process model, and rewriting the process model definition.

Lessons learned from the first application of the Cleanroom process in the SEL included :

- the fact that we could scale up to 30KLOC and use the process with changing requirements.

- The failure rate during test reduced to close to 50% over our typical project.

- There was a reduction in rework effort: 95% as opposed to 58% took less than 1 hour to fix.

- Only 26% of faults were found by both readers.

- Productivity increased by about 30%.

- The effort distribution changed in that there was more time spent in design and 50% of code time spent reading.

- Code appears in library later than normal and more like a step function than in the standard project.

- There was less computer use by a factor of 5.

We also learned that we needed better training for methods and techniques and better mechanisms for transferring code to testers. Testers need to add requirements for output analysis of code. As expected, there was no payoff in reliability modelling due to the inability to seed a model with such a small number of failures. One of the side effects of the project was that the Cleanroom development caused more emphasis on requirements analysis and the requirements writers were willing to refine their methods.

Based upon the success of this project, two new Cleanroom experiments were defined. A goal for both was to try to apply the formal models more effectively, i.e., use Mills' box structure approach. One project involved a change in the application domain, keeping the size of the project about the same (30KSLOC). The second project was a scale up to over 100KLOC and the use of contractors.

8.4 An Organization for Packaging Experience Models for Reuse

Improving the quality of the software process and product requires the continual accumulation of evaluated experiences (learning) in a form that can be effectively understood and modified (experience models) into a repository of integrated experience models (experience base) that can be accessed/modified to meet the needs of the current project (reuse) [BaRo91]. Systematic learning requires support for recording experience, and off-line generalizing, tailoring and formalizing of experience. Packaging requires a variety of models and formal notations that are tailorable, extendible, understandable, flexible and accessible. An effective experience base must contain accessible and integrated set of analyzed, synthesized, and packaged models that captures the local experiences. Systematic reuse requires support for using existing experience and on-line generalizing or tailoring of candidate experience.

This combination of ingredients requires an organizational structure that supports them. This includes: a software evolution model that supports reuse, processes for learning, packaging, and storing experience, and the integration of these two functions. We define a capability based organization to deal with such needs, which differentiates between the software development (Project Organization) and the packaging of experience in models (Experience Factory) [Ba89]. The Project Organization develops products with the support of reusable experiences from the Experience Factory tailored to its particular needs. The Experience Factory is a logical or physical organization that supports project developments by analyzing and synthesizing all kinds of experience, acting as a repository for such experience, and supplying that experience to various projects on demand. It packages experience by building informal, formal or schematized, and productized models and measures of various software processes, products, and other forms of knowledge via people, documents, and automated support.

8.5 Conclusions

The integration of the Improvement Paradigm, Goal/Question/Metric Paradigm, and Experience Factory Organization provides a framework for software engineering development, maintenance, and research. It takes advantage of the experimental nature of software engineering. It provide a framework for defining quality operationally relative to the project and the organization, a justification for selecting and tailoring the appropriate methods and tools for the project and the organization, a mechanism for evaluating the quality of the process and the product relative to the specific project goals, and a mechanism for improving the organization's ability to develop quality systems productively.

References

[Ba85a] V. R. Basili. *Quantitative Evaluation of Software Engineering Methodology* Proc. of the First Pan Pacific Computer Conference, Melbourne, Australia, September 1985 [also available as Technical Report, TR-1519, Dept. of Computer Science, University of Maryland, College Park, July 1985].

[Ba85b] V. R. Basili. *Can We Measure Software Technology: Lessons Learned from 8 Years of Trying* Proceedings of the Tenth Annual Software Engineering Workshop, NASA Goddard Space Flight Center, Greenbelt, MD, December 1985.

[Ba89] V. R. Basili. *Software Development: A Paradigm for the Future* Proceedings, 13th Annual International Computer Software & Applications Conference (COMPSAC), Keynote Address, Orlando, FL, September 1989

[BaPa85] V. R. Basili, N. M. Panlilio-Yap. *Finding Relationships Between Effort and Other Variables in the SEL* IEEE COMPSAC, October 1985.

[BaRo87] V. R. Basili, H. D. Rombach. *Tailoring the Software Process to Project Goals and Environments* Proc. of the Ninth International Conference on Software Engineering, Monterey, CA, March 30 - April 2, 1987, pp. 345-357.

[BaRo91] V. R. Basili, H. D. Rombach. *Support for Comprehensive Reuse* Software Engineering, IEE British Computer Society, September 1991.

[BaWe84] V. R. Basili, D. M. Weiss *A Methodology for Collecting Valid Software Engineering Data* IEEE Transactions on Software Engineering, vol. SE-10, no.6, November 1984, pp. 728-738.

[Dy82] M. Dyer. *Cleanroom Software Development Method* IBM Federal Systems Division, Bethesda, Maryland, October 14, 1982.

[Mc85] F. E. McGarry. *Recent SEL Studies* Proceedings of the Tenth Annual Software Engineering Workshop, NASA Goddard Space Flight Center, December 1985.

[SeBaBa87] R. W. Selby, Jr., V. R. Basili, and T. Baker. *CLEANROOM Software Development: An Empirical Evaluation* IEEE Transactions on Software Engineering, Vol. 13 no. 9, September, 1987, pp. 1027-1037.

9 Modelling Working Group Summary

Dan Craigen

9.1 Description

One of the central themes of formal methods is the mathematical modelling of digital systems. Furthermore, the new and evolving process of developing digital systems can use techniques derived from older engineering disciplines. Hence, there is interest in investigating how to model digital systems and how to model the digital engineering process. A clearer understanding of both classes of models should lead to increased quality and predictability of digital systems.

9.2 Discussion Topics

The following list of topics was distributed to the modelling working group participants prior to the workshop. At the time of distribution, it was made clear that the topics for discussion were open to change.

Topic 1: How does mathematical modelling help to increase our understanding of digital systems?

- What are the benefits of mathematical modelling? How do we achieve them?

- What are the risks of mathematical modelling? How do we avoid them?

- What are the limits of mathematical modelling? How do we identify them?

- What is required to validate that a mathematical model describes a digital system accurately?

- Identify general conclusions and recommendations.

Topic 2: What insights do other fields of engineering provide?

- Are we developing a new engineering discipline? If so, what are the primary defining characteristics of the discipline?

- Identify general conclusions and recommendations.

Topic 3: What can we do by 1995 with the mathematical modelling capabilities we have now?

- Are the current mathematical models sufficiently visible and explicit? If not, how is the problem rectified? Consider, for example, whether those models upon which we base our deductive reasoning are visible and explicit?

- What are reasonable goals for 1995? How do we measure progress toward them?

- What particular digital models are needed for critical applications? What ones exist now? What ones need to be developed? How do we get reusable models?

- Identify general conclusions and recommendations.

Topic 4: How do we expand the scope and effectiveness of our mathematical modelling capabilities?

- What are some challenging goals for the year 2000?

- What are the limits of our current mathematical capabilities to model digital systems accurately? How do we move beyond these limits?

- What additional models do we need to support critical applications beyond 1995?

- Identify general conclusions and recommendations.

Topic 5: How is modelling most effectively integrated into the system engineering process?

- What is the proper relationship between modelling and testing in the engineering process?

- How do these modelling capabilities integrate effectively into the engineering processes and tools of the future? What do we need to do now to ensure that this can happen in the future?

- What kinds of modelling other than mathematical can help enhance digital system safety?

- Identify general conclusions and recommendations.

9.3 Discussion

The remainder of this subsection is a report on the various topics actually discussed.

9.3.1 Why and to what end formal methods?

There was a general consensus that formal methods are one of our best hopes for developing trustworthy computer-controlled critical systems. It is only through mathematically-based techniques that we can achieve an absence of "surprises" (to the limits of our models) with fielded systems. There was also consensus that existing practices for developing software are terrible. Generally, it is the case that no one, including both vendors and customers, know what their systems will really do. This interminable state of affairs is a result of, at least, a poor understanding of how to handle the complexities involved with the systems we wish to create; and a software industry that is expanding so rapidly it has outgrown the pool of available trained talent.

Visualization of an engineered product is important for attempting to determine whether the product is "safe" to use. For example, a rotten wooden bridge or a rusted metal bridge can be identified through visual inspection as being unsafe. Formal methods can be thought of as part of a "visualization process" for software and hardware systems.

There is a need for mathematically-grounded techniques to aid in the development of computer-controlled systems. At the moment, many of the techniques that exist require significant mathematical expertise and an understanding of foundational issues. This is unlike other engineering disciplines where mathematically-grounded techniques are used, but full understanding of the underlying mathematics is not required—however, an understanding of the assumptions and limitations of the techniques is required. It was noted that "control theory" has evolved from a point where in-depth mathematical understanding was required to a point that engineers can now "push a button" and obtain the relevant answer.

The discussion on "Why and to what end formal methods?" ended with the conclusion that formal methods require a revolution in the development process, but *not* in the products being developed.

9.3.2 What is a formal method?

As is too often the case, the question of "What is a formal method?" arose. The group discussed a number of different connotations to the use of the term "formal." For example, the term has been used in

- contractual requirements: as "formal deliverables."

- design automation: as "formal*izing* the engineering process."

- engineering standards: as "a formal standard."

- mathematical modelling: as "formal*izing* a problem."

- automatic deduction: as "formal mathematics."

It was observed that "chip design" has a good set of tools (and underlying theory) leading to well-disciplined activities and constraints. However, the existing theory, primarily based on simulation tools, is nearing technological limits.

One of the goals of Computing Science research is the development of a reliable process from "requirements to product." We see formal methods as a thread through the entire process. Hence, we are attempting to establish an engineering process in which mathematics plays a role—but, not necessarily a central role.

9.4 How does mathematical modelling help to increase our understanding of digital systems?

The group listed the benefits of mathematical modelling as including the following:

- The models can be used to explore design alternatives.

- The models can be used for analysis of and expression of system behaviour.

- Different models will, in general, result in different perspectives of a system.

Having outlined some benefits accruing from mathematical modelling, various risks of mathematical modelling were identified:

- models may be inappropriate. For example, there may be an inappropriate selection of observables.

- models are incomplete. For example, no models of computer-controlled systems include the effects of α-particles.

9.4.1 What are the limits of mathematical modelling?

Having discussed the benefits and risks of mathematical modelling, the group then moved on to discuss the limits of mathematical modelling. The discussion led to various observations:

- Cost is often a deciding influence on how much effort should be directed at modelling.

- There are the usual limits on models of physical processes: they are *always* incomplete.

- Current formal methods based models are not integrated with performance models; nor with failure modes.

- There was consensus that our modelling techniques are in good shape with respect to sequential programs, simple hardware devices, and for limited forms of concurrency.

- There was consensus that our modelling techniques are in bad shape in that they are not being applied, they are not becoming widely known, there is little domain specific knowledge, and what seems simple for many individuals involved in formal methods, is viewed as being difficult by the masses.

- There are no good theories for abstract data types, unlike hardware, where cell libraries are widely used.

- One major complaint that arose during this discussion was that there is currently no traceability from requirements capture through to actual system implementation.

9.5 What is required to validate that a mathematical model describes a digital system accurately?

There was only limited discussion of this point. It was agreed that "accuracy" means the agreement between a model and physical reality. Moreover, it was concluded that testing and simulation may be used to validate models.

9.5.1 What insights do other fields of engineering provide?

Unlike other engineering disciplines, formal methods do not yet, in general, consider issues of fault tolerance and failure modes. This is one area of weakness that will need to be addressed if formal methods are to be widely used in the development of critical systems.

There was the observation that other engineering disciplines have a wide host of techniques that are used and discarded as necessary. The formal methods community, in particular, and computing scientists, in general, are still developing techniques that can be treated in a similar manner. With the newness of computing science, it will be some time before we reach the maturity that other, more established, engineering disciplines have reached.

In concept and aspiration, there is a high degree of commonality between formal methods and engineering. For example, both communities are interested in being systematic and developing "good practice."

9.5.2 What can we do by 1995 with the mathematical modelling capabilities we have now?

This discussion suggested that there is a lot to be achieved using existing capabilities. For example, there was agreement that:

- we can model hardware that corresponds to "interleaved sequential" behaviour. Where there are difficulties is with multiprocessor machines such as the IBM/PC. On the other hand, there seemed to be the view that modelling of a Cray/XMP was possible. Hence, there is no correlation between our ability to model hardware and the cost of the hardware.

- we can model parts of widely used programming languages. There are difficulties with the machine-dependent features of programming languages — as they require modelling of the underlying hardware.

- we can model various kinds of operating systems. These are difficult applications since they interact with the underlying hardware.

- we can develop reusable libraries of verified programs.

- we believe that, with existing technology, requirements can be tracked throughout the development process to implementation.

9.5.3 What is our target for the year 2000?

The main target for the year 2000 is to develop an engineering process that will allow for the development of highly automated systems that have no surprises to the designer, nor to the user. We are striving for predictability of system behaviour.

9.5.4 Formal methods and safety standards

For a part of the discussions, AVM Mike Brown was invited to join the group and to present his general opinions about formal methods. AVM Brown, now retired, was one of the main U.K. government forces behind the development and acceptance of MoD standards 00-55 and 00-56.

AVM Brown observed that the terminology of formal methods makes it difficult for management individuals (with government and industry) to understand and accept. Specifically, questions such as "What are we modelling?" and "How can we model something that doesn't exist yet?" appear to be stumbling blocks. Additionally, if management is told that programming is primarily symbol manipulation and that models are better than implementations, why not throw away the implementations? Don Good argued that symbol manipulation is not the correct perspective and that with formal methods we are modelling physical reality in terms of the behaviour of silicon. Good, and others around the table, could not understand why modelling terminology is acceptable in other disciplines, but not with formal methods. When this part of the discussion was reported to the plenary session, Jean-Raymond Abrial commented that, perhaps, the use of the term "blueprint," rather than "model," would be preferable.

AVM Brown also discussed some of the difficulties the government has in writing procurements. For example, requirements must not lead to a bias in procurement. In his view, models help in requirements capture and potentially help to alleviate the problems with bias. It was further noted that "good commercial practice" should use modelling ideas—modelling is "best practice" and the customer should not have to pay extra for a company to use modelling techniques.

Finally, it was observed that there is little prestige in identifying oneself as a "software engineer." The reward for a good software engineer is to be "promoted" to management. In addition, major software engineering decisions are often made by junior individuals; this is not the case in other disciplines.

9.6 Conclusions

In the final wrap-up session a few general conclusions arose:

- Formal methods are good where there has been considerable experience with the application domain. Specifically, CICS (IBM, Hursley) and the oscilloscope work of Textronix demonstrated how useful formal methods could be when extensive experience and understanding of the domain existed.

- Thinking about the entire system, rather than the computer-controlled components solely, seemed to help the development process.

- Formal methods have the potential of being time and cost reducers for getting a system through the development process and out to market.

- Often, numerous models of a system are necessary. Different perspectives lead to better understanding of system behaviour.

- Finally, it was concluded that the formal methods community has substantial technology to offer to developers, but that there are serious problems with transition. A substantial education effort and successful demonstrations of the technology are required.

10 Quality Assurance Working Group

Richard A. Kemmerer

10.1 Group Description

The main objective of the workshop was to accelerate the diffusion of formal methods into practice. It is the belief of most Formal Methodists that although formal methods can enhance some traditional quality assurance techniques, they are not a substitute for these techniques. That is, the use of formal methods cannot completely replace traditional assurance techniques. Therefore, within the framework of FM'91 the primary goal of the Quality Assurance working group was to determine how formal methods can complement established quality assurance techniques so as to yield high levels of assurance of intended behavior.

The following ideas were circulated to the group members prior to FM'91 to stimulate thought.

1. To achieve the group's goal it is necessary to identify the traditional quality assurance techniques that are being used today.

2. It is also necessary to identify examples of where formal methods have already been integrated into traditional quality assurance techniques. The success of each of these experiences should be determined.

   - was the system better as a result?
   - would the practitioners use this approach again?
   - what would be done differently if this approach were to be used again?

3. For each of the traditional techniques identified in 1 for which there are no examples of integrating formal methods into the technique, are there ways that this might be achieved or fundamental reasons why it is not feasible? Also, are there other ways of integrating formal methods into those quality assurance techniques that already contain some formalism (that is those that were discussed in 2).

4. A fundamental question that will have to be answered by the group is what is quality assurance and how can one measure it?

5. Are some formal method approaches and tools more amenable to integration into the quality assurance process? If so why?

6. How can one infuse the group's ideas for integrating formal methods into the quality assurance process into the minds of the software practitioners? That is, how can one get the software houses that are building large software systems to accept the integrated approaches as part of their standard quality assurance process.

10.2   Quality Assurance vs. Quality Control

In the plenary session before the working groups met Vic Basili gave a talk "Software Quality: A Modelling and Measurement View." Motivated by the definitions that Vic gave for quality assurance and quality control the first question that arose was whether the group was interested in quality assurance, as the name implied, or in quality control?

Quality control is what one does to produce better software, while *quality assurance* is how one convinces someone else that the software is better. Thus, quality control is internal to the project and is product oriented. In contrast, quality assurance is external to the project and is process oriented.

In traditional manufacturing there is an implicit link between process and product. That is, monitoring the manufacturing process is usually sufficient to assure a good product (Of course, one must also start with quality raw materials, etc.). Thus, one develops a manufacturing process that includes quality control, which in turn assures a good product. This *does not* work for software. With software development one does not develop a process - development is the process.

As a result of this introspection the group decided that what it was concerned with was quality control.

10.3   What is a Formal Method?

After having decide that it was interested in quality control, the group next asked what could be done that had a good chance of being used to integrate formal methods into the quality control process. The suggestion that got the most favorable response was to teach people how to read (as in the cleanroom approach). This was supported by the fact that one needs to know how to read well to know how to write well. In addition, part of learning how to read is learning how to abstract, and the group argued as to whether abstraction was a "formal method." This led to a long discussion on what constituted a formal method.

After much discussion, which was similar to the discussions reported in the proceedings of the FM'89 workshop, the group agreed that there were at least three different definitions of a formal method being used in the group. These three definitions were denoted as FM1 FM2, and FM3.

FM1 - a formal method is a mathematical model of a software product.

FM2 - a formal method is a mathematical model of a software object, where object includes product, process, resources, etc.

FM3 - a formal method is a set of structured principles about which one can reason.

After analyzing these definitions the group concluded that one of the reasons that there was so much difficulty in defining a formal method is that there are a whole range of definitions. However, these definitions seemed to be dependent on the axis: the degree of formality and rigor and the scope to which this formality applied. For instance, definition FM2 considers a larger scope (i.e., it includes process, resources, etc. for analysis) than definition FM1, which considers only the product itself. Both can have the same level of formality or rigor, however. In contrast, definition FM3 differs from FM2 in the degree of formality while the scope may be the same. Thus, FM3 would permit cleanroom while FM2 would not.

10.4 Integration of Formal Methods and Quality Control

The group next decided to address some of the original questions that were posed for this group. However, before being able to discuss what the traditional quality control techniques are it was necessary to discuss what software qualities are. The group compiled the following to be a partial list of interesting qualities: correctness, readability, maintainability, efficiency, reliability, cost effectiveness, usability, functionality, ability to find defects.[5]

It was decided that formal methods can have a direct effect on a large number of product qualities. Some qualities, such as correctness and maintainability, are directly affected, while others would be affected as a result of building models and analyzing them (e.g., running simulations).

The group next decided to identify examples of how formal methods and established quality control techniques could complement each other. The most obvious is the idea of testing formal specifications. Another is the use of formal specifications to generate more detailed specifications and design test cases, as well as test cases for the code itself.

A formal model can also be used as an oracle. That is, it can be used as confidence builder for testing (e.g., when the exact correct output is not known). Finally, if the formal specification is executable, it can be used as a rapid prototype.

The group agreed that although there are many techniques with possibilities, for the most part these techniques have not been exercised on real projects.

10.5 A Plan to Integrate Formal Methods into QC/QA

The final thing that the group did was to outline what approach might be taken if one were tasked to implement a formal methods-based quality control and quality assurance department in three years using two full-time persons. The transfer outlined is a four phase project, with an additional prephase called the conditioning phase. During the conditioning phase plans are made for carrying out the transformation. This is the time when one gets management to buy-in to the approach, appropriate projects and personnel are selected, and baseline data and process metrics are collected.

10.5.1 Phase 1

In the first official phase, Phase 1, the basics of reading and writing are introduced. This includes the reading (writing) of requirements, specifications, designs, and code. These are read (written) with particular attention paid

---

[5] It was noted that a previous study of software qualities had identified more than 70 qualities.

to correctness, testability, traceability, etc. A case study, probably from the business community, will be used during this phase.

In Phase 1 structural principles, abstraction, error inspection, and teamwork will be defined and taught. In addition, error data collection, testing (including test case generation, use of assertions, and input for static and dynamic analysis), documentation technology, and configuration management technology will be established.

This phase is for technical personnel and management personnel.

10.5.2   Phase 2

Phase 2 is the structuring and reasoning phase. During this phase some methodology such as Mill's cleanroom, Z, or OBJ will be introduced. Mathematical structures, such as invariants and input/output assertions are introduced, and mathematical reasoning using abstraction and correctness checking are introduced. During this phase the reading techniques are reinforced and refined, but the emphasis is on correctness. This phase will also see improved error data collection, test data generation, and documentation.

This phase is also for technical personnel and management personnel.

10.5.3   Phase 3

This is the process definition phase. Different process models are introduced. Informal, semi-formal, and formal techniques are defined, and for each technique when and how to use the technique, what qualities are addressed, what tools are available to support the technique, and issues of how to measure the technique's effectiveness are described.

This phase is for management and senior technical personnel.

10.5.4   Phase 4

This final phase is called the advanced phase. Advanced tools, such as testers, provers, and code generators are introduced. Advanced proof and error detection techniques are used, and rigorous types of analysis are performed. A large systematic case study will be analyzed during this phase.

11   Design Methods Working Group

Chris Sennett

11.1   Description

The aims of this working group were to look at existing systems engineering design methods and establish how formal methods could be integrated into such practice. Suggested issues to be discussed by this working group included:

- Which formal methods are best suited to what stages of development and what kinds of development?

- What exactly are the expected benefits of such integrations?

- Do we already have examples of such integrations?

- Can we characterize the technical gains from each integration?

- Such a process might ease the transition of formal methods into practice. How can we best capitalize on this?

- What effect will such an integration have on formal methods and existing tools?

- What should we be doing to advance the process?

11.2   The context of formal methods

The working group made comparisons between the state of the art as reported in FM'89 and that current, so it is useful to give some of these conclusions which set a context for the rest of the discussion. Probably the most important of the differences was that whereas FM'89 had been oriented towards achieving trustworthy software for critical systems, the emphasis now was much more on the use of formal methods to support software development in general. This was reinforced by the increasing number of practical applications it was now possible to cite. Whereas in FM'89 the discussion had been on academic research work and industrial work in progress, in FM'91 the atmosphere was much more of formal methods providing cost effective capabilities, supported by examples of completed projects and marketable products.

A survey of the working group members produced a number of examples. Anthony Hall's talk gave an example of the use of formal methods by Praxis where the benefit could be categorized as quality and visibility. Jim Woodcock mentioned the use of analysis and proof in the development of Transputer chips by Inmos (described in [1] and [2]). The use of Z in the development and maintenance of CICS [3] is relatively well known and interesting as the application of formal methods to a large scale system, by a large scale concern. A particularly interesting example was given by Garlan describing the use of formal specification to capture the intellectual assets represented by the communal knowledge of system designers of the architecture of a product range.

The particular experience discussed was in the design of oscilloscopes by Tektronix, resulting in a quicker time to market for a range of products. Some technical aspects of this work are described in [4], [5], but again the interest is in the fact that in the industrial context, formal methods have an application which is not the traditional verification of correctness. Even within that field of achieving certifiable software, the application of formal methods does not follow the traditional verification and refinement ideas. For security for example, where much of the certification activity has taken place, the emphasis is on the verification of properties, rather than functional correctness. This distinction is discussed in [6].

Thus even a small sample of experts in the field were able to provide a number of realistic examples of the use of formal methods in industry. Before going on to discuss the implications of this, the group's views on the general technical capabilities of current formal techniques will be given covering the topics of proof and assurance; analytical and modelling capabilities; and foundational and semantic aspects.

11.3 The role of proof in assurance

This has proved to be one of the most contraversial aspects for the use of formal methods and the topic gave rise to lively discussions within the working group. The point at issue is the extent to which proof adds to assurance. For example, in the case of a microprocessor used for safety critical applications, is it better to choose a formally verified chip or one which has been widely used. The problem is that the question is almost meaningless without being more precise as to what has been formally verified, what the safety critical concerns actually are, and what widely used actually means. Even with this precision, two qualitatively different aspects of assurance are being compared so the comparison cannot really be regarded as legitimate.

Nevertheless, the question keeps on being asked because few formal verifications have been attempted for real-life situations. As a result, verification has no track record of reliability: the only reasons for believing it adds to assurance are theoretical ones and it is not obvious that the techniques, when employed in the real industrial world, live up to the expectations of them. The word proof has connotations of ultimate truth in the minds of the general public, which would not be shared by a mathematician. Researchers make few claims for the assurance delivered by a formal proof, apart from the fact that it does substantially add to assurance. This of course lays an obligation on researchers to make clear the significance of proof technology, but it also has an impact on design methods when used with formal proof.

The best way of presenting the assurance that proof gives seems to be to make it clear that proof is always about a model of the real system. So there are two aspects to assurance, the quality of the proof as a symbolic function on the model and the relevance of the model to the real world. With regard to the first aspect, the peculiar nature of both software and hardware proofs is their size: proofs of a million formal steps are not unknown, and while one would normally break this down and give it some structure, proving the correctness of quite small programs involves dealing with sub-proofs in quantity. The consequence is that tool support is essential if the proof itself is to contribute to assurance. An informal proof has value in giving a precise explanation of the conditions for correctness, which certainly adds to assurance, but the proof itself could not be trusted if the number of formal steps was large. The proof tool itself will need some measure of assurance. This is probably somewhat short of formal verification of correctness, but it ought at least to be possible to state the condition for the tool not to prove a false theorem.

Of more significance to assurance is the aspect of relevance of the proof. This is concerned first of all with the validity of the abstraction used to form the model and then with the significance of the theorem proved in relation to the properties desired. For validity one is concerned not only with the faithfulness of the properties being modelled but also with the interpretation of the abstraction. For example, a proof about the behaviour of a system in, say, CSP, would be concerned not only with whether the model was valid (time irrelevant say, or deadlock not a problem in the trace model) but also with whether the process representation was actually found in the implementation language or the operational code. For significance, a typical question in the security field would be whether a proof of information flow properties represented all that was of significance for security.

Relevance of a proof can only be a human judgement. The consequence for proof tools is that presentation is of overriding importance. If the theorem to be proved is lost in jumble of symbols, its proof does not add to assurance. Of the three factors relevant to proof in design methods, namely automation, trustworthiness and presentation of the proof, the last is probably the most important.

11.4 Analytical capabilities of formal methods

Analysis is the term used here for all those ways of processing formal texts, short of formal proof. Essentially, a formal text is either a specification or an implementation and the problem with formal specifications as opposed to implementations is that they cannot be executed. Consequently it is hard to tell whether a specification of any size is meaningful, consistent or has other attributes of quality. A program on the other hand is compiled, which usually means being checked for type correctness, and then executed, so that it may be tested. Both of these give primitive but effective methods for acceptance of programs and analysis of proof is intended to provide a similar means for accepting specifications.

The standard analytical method for a specification language is syntax and type checking. This is practised widely, is low-cost and effective. This can be supplemented by informal reasoning to strengthen considerably a specification written in natural language and diagrammatic notations.

Apart from this, few other forms of analysis are applied to specifications. There are however, a number of possibilities. First of all there is the question of consistency. The general problem of establishing the consistency of a specification is beyond the state of the art, but it is certainly possible to generate verification conditions which if proved would ensure consistency. The second possibility for analysis is the display of the structure of a specification using structure diagrams as with implementation languages. Thirdly there is the possibility for symbolic execution and animation which would enable properties of a specification to be tested. Fourthly there is the possibility of relating an implementation to a specification and generating the conditions for compliance.

Work has been undertaken under each of these headings, but apart from the last possibility has not yet resulted in widely available tools.

Analysis of implementations is relatively well established, often using the method of abstract interpretations on the implementation language regarded as an algebra. Techniques available in this field include compiler optimizations, test coverage analysis and information flow and data use.

11.5   Foundational capabilities of formal methods

The oldest application of formal notations has been in the definition of programming languages. The most successful application of a mathematical technique which is essentially formal, is in the specification of the syntax of programming languages and the derivation from the syntax of analysers capable of recognising the sentences of the language. Formal semantic definitions of languages have been much less successful. This is largely because of their complexity, the extreme example being the draft formal semantics of Ada [7]. The semantic definition can be regarded either as a demonstration that the language has a meaning or as a precise explanation of the informal definition or as a specification of a compiler. Complexity militates against all of these objectives with the result that semantic definitions have not been as useful as expected. Some parts of the formal methods community have reacted to this by taking the view that implementation languages should be simplified: this is not entirely special pleading and it seems reasonable to argue that for safety critical work one should work with languages which are capable of having formal semantics given to them. These need not necessarily be toy languages as shown by the examples of Verdi [8] or standard ML [9].

The complexity issue has not been entirely satisfactorily resolved. A structure for language definition in the form of static and dynamic semantics is often used and this is a considerable help, but it would be useful to have more structure still which could be related to typical compiling functions.

Implementation languages are not the only languages associated with software development and it is useful to be able to give formal definitions of their meaning. The most common notations employed in software design are the diagrammatic notations associated with the structured methods. Some, such as JSD, are actually based on ideas used in process algebras, so it would be relatively straightforward to give formal semantics to them. Others, such as the data flow diagrams used by a number of methods, have a rather imprecise meaning and the formalization presents many problems. These are indicative of the fruitful nature of the interaction between structured and formal methods: attempts to formalize data flow diagrams, for example [10], can do so at the expense of fixing on one interpretation of the diagram. The fact that it can be interpreted in more than one way may be a drawback or may be one of the advantages of the diagram, but it is certainly true that the interpretation should be made clear for each application and the formalization forces one to do this.

The formalization of data flow diagrams also calls into question the motivation for providing a semantics. The diagram is supposed to explain the implementation, so there must be some relationship between the meaning of the diagram and the implementation. However, the intuition associated with a diagram is often concerned with information rather than data, so there may be meanings associated with the diagram, concerned with the interpretation of the data (for example, this data is a command, the other is a message) which are not present in the implementation, as well as information in the implementation (for example, access control in security) which is not present in the diagram. This is in addition to the process versus sequential interpretation of the boxes in the diagrams.

Data flow diagrams are used for more than one purpose and for some purposes, they are not entirely satisfactory. Formalization within certain contexts is likely to produce new diagrams which may meet the purposes of the design methodology better. In particular it seems that the data flow diagram is mainly used as an abstraction of the implementation rather than a specification for it, so it should be designed for the customer rather than for the programmer. The particular view taken will affect the form of semantics chosen and the role of tools to support the diagram.

Other diagrams and notations, particularly at the architectural level, will almost certainly be required in the future. Unfortunately, formal methods practioners have a penchant for introducing new notations almost equalled by programmers' reluctance to use them. At least the use of formal semantics ensures that notations are precise and unambiguous. Acceptability can be ensured by matching the notation to the needs of the users, which is the topic of the next section.

11.6 The role of formal methods with respect to the software development process

Rather than define a particular development process model, two phases only will be considered, namely the process of requirement formation leading up to the procurement specifications, and the process of design and implementation from a specification. Within these phases a method may offer the following qualities:

- Visibility - an understanding of the proposed system or its implementation.

- Validation - a demonstration of the fitness for purpose of the proposed system or the implementation.

- Compliance - a method of ensuring that the implemented system will meet the requirement.

For each of these qualities there is a third dimension of properties such as performance, security, availability and so on which will be specific to a given application. A given formal technique may be applicable to a given triple of (phase, quality, property) or may be more general. The working group did not consider all the elements of this matrix, although in principle it indicates the criteria for judging the application of particular formal methods. Instead some of the particular areas where recommendations can be made were highlighted.

With regard to the requirement formation phase, there are few existing informal methods and the formal methods have been mainly devoted to security requirements. The working group was quite clear on the conclusion that high level specifications were important in the formation of requirements, but that much work remained to be done to elevate this expectation to practice. In this respect the work on the specification of architectures for Tektronix referred to earlier is particularly relevant. It was noted that this work was undertaken in the context of product development, rather than bespoke system development as would be normal for government procurements. The product development context is particularly suitable for this application: the range of products allows for a flexible development schedule; the common ownership of the software for the entire range allows for interaction between product engineers and maximises the opportunities for re-use of software and ideas.

Bespoke systems on the other hand are independently and competitively procured, all too often starting from scratch. There is considerable scope for building up a domain specific infrastructure in which formal notations could play a substantial part. An initiative in this area must be customer led, but will need to be collaboratively undertaken with industry. It requires some change of emphasis in procurement policy, but above all, a clear lead from the customer.

Apart from the benefits in clarifying the common requirements in a range of procurements, there are two other factors which would tend towards this approach to procurement. The first is the existing investment in systems, the need to enhance and interconnect them and to evolve the capabilities they provide. This will counter the tendency towards procurements from scratch and make it very desirable to reverse engineer the existing systems towards achitectural specifications which can be used as a basis for enhancement. The other factor is the increasing dependence on standards and commercial off-the-shelf software. In order to be able to use these, it will be essential that they have not only well defined interfaces, but also well defined profiles of use.

11.7 Formal methods during development and implementation

This phase is supported by many formal and informal methods; interaction between the two styles is taking place and can only be beneficial. However it was felt quite strongly that progress would be made by understanding the role of the methods within the software development process. It is an indication of the maturity of the field that this is much more of a concern now than it was a few years ago, but it needs to be constantly borne in mind that a method must have a purpose and meet some need within the development process if it is to be useful. It is clear that many methods will be beneficial within the process: there is a need for interaction between methods, but the tendency to produce a uniform notation for all aspects of software development should be resisted. Specification languages should not be enhanced to include implementation language features: neither should implementation languages be enhanced to include specification features. Instead each method should be used within its own terms and the requirements for interaction between the methods should be carefully thought out. In many cases simple consistency checks between notations may be all that is required, rather than a rigorous translation, refinement or abstraction.

Formal methods are potentially revolutionary, both in the concepts required and benefits provided, but revolution is counter productive to successful software management. It is clear that progress in formal methods will be obtained by making methods supplement existing methods, rather than replacing them: this is the only way to guarantee an increase in assurance coming from their use.

Management changes are, however, necessary. Formal methods increase the quality of the product, rather than its quantity. Formal methods support progress by documents and understanding, rather than lines of code. Formally specified code is often shorter and faster, simply because it is well-structured and well understood. It is these objectives of quality which management must pursue, rather than mere lines of code.

A particular recommendation of the working group in this area was that progress in developing appropriate formal techniques would be considerably assisted by a better understanding of the issues which arise in real life software developments. The presence of realistic project analyses, freely available to the researchers would orient the research appropriately. A further recommendation was an emphasis on the importance of case studies. Notations and theory already existing could meet many needs, it simply needs some fairly large scale examples working out to illustrate how existing techniques apply.

11.8 Education

As with FM'89 there was a call for more education. This is needed at four levels: foundations, engineering, management and practioners. The first two of these are addressed in universities and schools, the other two in industry. It is quite clear that in this field, activity in education is strongly customer led. If employers demand experience of formal methods, formal methods courses will be demanded by students and there will be no major problem in providing them. Without this there will be no demand by the students for the vocationally oriented courses. This applies with even more force to industry: without a demand for formal methods courses, there will not be a supply. As the techniques are used within industry and the benefits realised the demand will surely build up, but because the benefits are in improved quality, the build up will be slow without customer pressure.

11.9 Tools

Tools continue to be an issue in the industrial take-up of formal methods. There is a clear need for tools to support the integration of structured and formal methods, for specification language tools to take more account of configuration control and for integration of tools with compiling systems, particularly for high integrity. Tool development is a high risk, low profit activity in itself, the main profit coming from associated consultancy and the training in use which is a very necessary part of a practical tool. Tool development would be fostered by standardisation, although apart from specification languages this may be a little premature. Apart from this, there is scope for research in demonstrating tools that adequately support the software development process.

References

[1] Barrett G. Formal methods applied to a floating point number system. IEEE Transactions on Software Engineering 15(5):611 - 621, May 1989.

[2] Inmos Ltd. Specification of the instruction set & Specification of floating point unit instructions. Transputer Instruction Set - A Compiler Writers Guide, page 127 - 161. Prentice Hall, Hemel Hempstead, Herts, UK, 1988.

[3] King S. The CICS application programming interface: program control. IBM Technical Report TR12.302, IBM United Kingdom Laboratories Ltd., Hursley Park, Winchester, Hampshire SI21 2JN, UK, December 1990.

[4] Spivey J M. Specifying a real-time kernel. IEEE Software pp 21 - 28, September 1990.

[5] Garlan D and Delise N. Formal specification as reusable frameworks. In Bjorner, Hoare and Langmaack (eds) *VDM and Z - Formal Methods in Software Development*, vol 428 of Lecture Notes in Computer Science, pages 150 - 163, Springer Verlag, 1990.

[6] Arthan R D. On formal specification of a proof tool. In Prehn and Toetenel (eds) *VDM '91 - Formal Software Development Methods*, vol 551 of Lecture Notes in Computer Science, pages 356 - 370, Springer Verlag, 1991.

[7] Commission of the European Communities. The draft formal definition of Ada. Commission of the European Communities, 1986.

[8] Craigen D, Kromodimoeljo S, Meisels I, Pase B and Saaltink M. EVES: an overview. In Prehn and Toetenel (eds) *VDM '91 - Formal Software Development Methods*, vol 551 of Lecture Notes in Computer Science, pp 398 - 405, Springer - Verlag, 1991.

[9] Harper R, Milner R and Tofte M. The semantics of standard ML. Laboratory for the Foundations of Computer Science report ECS-LFCS-87-36. Department of Computer Science, University of Edinburgh, The Kings Buildings, Edinburgh, EH9 3JZ, Scotland.

[10] Randell G P. Translating data flow diagrams into Z (and vica versa). RSRE report 90019. Defence Research Agency, Malvern, UK, 1990.

12 Conclusions

It is clear that formal methods continue to make progress at a healthy pace. Significant advances have been made on both sides of the Atlantic since FM'89. The formal methods community's self-perception has greatly matured and there is now a far clearer consensus as to the scope and limitations of the discipline. The danger of overhyping the technology is clearly recognized by the practitioners if not by all on the fringe. A more realistic perception of the role of formal methods with respect to existing practice has also emerged. It is clear that formal methods are not a substitute for such well established technologies as quality assurance and testing but that rather formal methods should be integrated with such existing practice. For example, testing can be used to add credence to assumptions underlying a formal analysis. Conversely, formal methods can guide the development of a testing strategy.

A major conclusion to emerge was the conviction that formal methods are now ripe for technology transfer. The methods and tools are now reaching the stage at which they can be used effectively on an industrial scale. Admittedly great care must be taken in the choice of methods and tools for any particular application. Care must also be taken to select the appropriate stage ar stages of the development at which to apply them. Thought must go into the question of how far to push the techniques. For example do you just do some formal specification? Do you even attempt any proofs? If so what proofs? By hand or by machine? What assumptions do you choose to leave unproven?

Thus, far from being a sledge-hammer with which to demolish high assurance applications, formal methods are a large and varied tool-bag. There is great flexibility with which can be exploited to achieve the best, most cost-effective results. What is still lacking, is a clear body of knowledge and advice on finding your way around this tool-bag. This is a major and valuable task waiting to be addressed.

Principal conclusions were that tools and training are both areas of weakness and in need of special attention. Without these in good shape, formal methods would have little hope of really taking off in the industrial context.

The perception of formal methods from outside, for example from funding agencies and project managers, was improving but still not very healthy. More effort to publicize successes, and indeed to achieve successes, is needed. All the time taking care to keep all claims realistic and honest.

The styles and approaches on either side of the Atlantic differed significantly. Far from being a disadvantage this should be turned to good use. The best aspects of both should be drawn out and exploited. Stimulus should be provided for more interaction between researchers and developers.

A subtle but significant shift of perception emerged from the discussions, namely that the achieving of quality systems is process oriented not product oriented. This seems obvious in retrospect but some of the early applications of formal methods tended to fall into the error of applying them in a rather retrospective fashion. The results tended to be expensive and of dubious relevance. The techniques must be applied during the development not just to the final product.

A   Survey of Formal Methods Tools

The ABLE Project

Participants (project leader, group members):
David Garlan, Robert Allen, Curtis Scott, Robert Nord, John Ockerbloom

Survey contact:
David Garlan
School of Computer Science
Carnegie Mellon University
5000 Forbes Avenue
Pittsburgh, PA 15213
USA

Level of effort (person-years, duration):

Description:
Academic research project to investigate the use of formal methods for describing and analyzing software architecture.

Accomplishments:
Developed several formal models of common software architectures.

Published articles or reports:

- *Formalizing Design Spaces: Implicit Invocation Mechanisms.* David Garlan and David Notkin in Proceedings of "VDM'91: Formal Software Development Methods" October 1991. Pages 31-44, Springer-Verlag, LNCS 551.

- *A Formal Approach to Software Architectures*, Robert Allen and David Garlan. January, 1992. Submitted for publication.

Status:
Active

Start date:
September 1991

Completion:
continuing

Future developments:
Building an environment, called Aesop, to support development of formal models of software architectures.

Strengths, weaknesses and suitability:
Currently only a research prototype

External users:
None yet.

Applications:
Formal methods in software engineering.

Availability:
Eventually to be distributed.

Additional remarks:
none

Ariel

Participants (project leader, group members):
Mark Bickford, Steve Brackin, Bret Hartman, Doug Hoover, Mark Howard, James Morris, Garrel Pottinger, Sanjiva Prasad, Ian Sutherland

Survey contact:
Bret Hartman
ORA Corporation
301A Dates Drive
Ithaca, NY USA 14850-1313

Phone: (607) 277-2020
FAX: (607) 277-3206
Internet: bret@oracorp.com

Level of effort (person-years, duration):
4 man-years per year for the last 4 years.

Description:
The Ariel project focuses on investigating advanced methods for formally proving real number properties in programs that use machine real arithmetic. To support proofs of mathematical software containing floating point computations, we are using the theory of asymptotic correctness. Asymptotic correctness allows proofs of mathematical software to be independent of the precision of a particular machine.

The Ariel project is part of DARPA's Ada STARS program. We are developing methods for proving the correctness of mathematical software and supporting numeric error exceptions in Ada programs. Additionally, we are studying methods for demonstrating Ada program termination. These areas of study complement the capabilities provided by ORA's Penelope verification environment (another Ada STARS project).

Accomplishments:
Early in the Ariel project, we produced a formal semantical model of asymptotic correctness and a prototype verification system to verify programs that are written in a subset of the C programming language.

Support for the verification of real number properties has now been integrated into the Penelope system. The combination of Penelope and Ariel create a cohesive verification environment for proving the correctness of a variety of Ada programs.

Published articles or reports:

- Mark Howard, *Verifying Asymptotic Correctness*, Proceedings of the Fourth Annual Conference on Computer Assurance (COMPASS '89),

June 1989.

- ORA Corporation, *Formal Verification of Mathematical Software*, Technical Report RADC-TR-90-53, Rome Air Development Center, May 1990.
- Sanjiva Prasad, *Verification of Numerical Programs Using Penelope /Ariel*, ORA Corporation Technical Report, January 1992.

Status:
A prototype verification environment exists that supports mathematical software written in Ada.

Start date:
May 1986.

Completion:
September 1992.

Future developments:
Based on the application of Penelope/Ariel to real number programs, proof capabilities will be improved and refined.

Strengths, weaknesses and suitability:

Strengths Proofs of real number properties are integrated with other correctness properties within an interactive proof environment.

Weaknesses There is still minimal support for demonstrating that a program computes a result within specific error bounds. We plan to improve numeric error analysis in later releases.

External users:
None until the system is exercised further within ORA.

Applications:
Formal proofs of high assurance mathematical software in Ada.

Availability:
Ariel will be available as part of the standard Penelope release in mid-1992.

Additional remarks:

Ada Verification Interface (AVI) – lightweight verification tools

Participants (project leader, group members):
Doug Weber, Cheryl Barbasch, Steve Brackin

Survey contact:
Cheryl Barbasch
ORA Corporation
301A Dates Drive
Ithaca, NY USA 14850-1313

Phone: (607) 277-2020

FAX: (607) 277-3206
Internet: cbarb@oracorp.com

Level of effort (person-years, duration):
2 person-years

Description:
The AVI project is a subtask of the STARS (Software Technology for Adaptable, Reliable Systems) Software Engineering Environment (SEE). The objective of STARS Task 18 is to develop methods for the specification and verification of secure and trusted systems implemented in Ada, and to build tools within the STARS SEE that support these verification methods. The methods developed must include, at a minimum, a mathematical theory for verifying that an algorithm implemented by a given Ada program meets its specification. The tools must allow a programmer to apply this theory conveniently to prove specified properties of Ada programs.

Subtask US18.1.2 consists of the Ada Verification Interface tools. Specialized, or "lightweight", verification tools check for the presence, or verify the absence, of commonly occurring programming errors. The types of errors of interest are run-time errors that are not normally detected at compile time by an Ada compiler. Examples include: erroneous executions; incorrect order dependences; raising of predefined exceptions, such as CONSTRAINT ERROR; infinite loops; and deadlock. General verification systems such as Penelope are typically used to prove the absence of these errors. In contrast, the aim of specialized verification tools is to automate the analysis of these properties, eliminating the need to construct proofs.

We have identified a small number of run-time error conditions for which verification techniques exist and have produced prototype implementations of two specialized tools: one designed to check for the presence, or verify the absence, of erroneous executions or incorrect order dependence due to aliasing of parameters to subprogram calls, and one to check for definedness of variables.

For the initial prototypes we were required by STARS to use the IRIS-Ada toolset funded under STARS Task US20 to provide an internal semantic representation of Ada source code for the verification analysis. Since IRIS is no longer being funded we are in the process of converting to ASIS (Ada Semantic Interface Specification), which is a vendor-independent non-proprietary bridge between Ada tools and Ada libraries.

Accomplishments:
Two prototype tools are presently implemented.

The tool, check_alias, checks all parameters of subprogram calls and generic instantiations of an ada unit for incorrect order dependence, and checks all subprogram calls for erroneous execution due to aliasing of subprogram parameters. It prints diagnostic information, much like LINT for C programs. If there are no error messages printed then the Ada unit is guaranteed to satisfy the specification:

Neither choice of parameter passing mechanism nor order of parameter evaluation by the compiler can affect the visible behavior of the program during execution (assuming the call is not abandoned).

The tool, check-def, conservatively checks a restricted subset of Ada programs for definedness of variables. It constructs a dependency call graph and legal compilation order (if possible) for an Ada program, then creates a control flow graph for sequences of program statements. It follows the control flow and uses data flow analysis to analyze definedness of each scalar variable in the order it appears, working upward in the program's dependency graph. It prints results of the analysis for each variable. In the present incomplete implementation, it only partially analyzes programs and does not yet handle non-scalar variables, recursion, generics or concurrency.

Published articles or reports:

- ORA Corporation, *Ada Formal Methods in the STARS Environment, STARS Task US18*, Technical Report submitted to STARS, December 18, 1990.

- ORA Corporation, *Specialized Verification Tools in the STARS SEE*, Technical Report submitted to STARS, August 20, 1991.

Status:

The tools depend heavily on the IRIS-Ada translation tools provided by Incremental Systems. There are significant bugs in the analysis performed by the IRIS tool, and additional development is not likely to be continued. To run the check-alias or check-def tools, Incremental's Ada-to-IRIS executable must be run first.

The check-alias tool presently checks all subprocedure calls and generic instantiations, but does not yet handle Ada units with tasks. The check-def tool presently has several areas that are not handled, including: recursion, concurrency, generics, programs with non-scalar variables, some statement types, and analysis across subprogram call boundaries.

We are just beginning to convert the tools to ASIS, using the Telesoft compiler. The big advantage is that the Ada units being analyzed are in an intermediate form produced by the compiler itself, and other compiler vendors are planning to provide an ASIS interface in the near future.

Start date:
July 1990.

Completion:
September 1992.

Future developments:

Convert the existing tools to ASIS, and provide other verification tools: predicate transformation, constraint checking, elaboration-order checking, and termination.

Strengths, weaknesses and suitability:

Strengths The tools are intended to analyze at compile-time any Ada program, not just "toy" programs, to guarantee the absence of certain types of run-time errors that are typically not checked by compilers. The set of programs they can handle in the present implementation is restricted to the subset of Ada that an Ada-to-IRIS front-end can parse correctly.

Weaknesses The tools presently depend on a front-end analysis tool that was never completed and is not likely to be supported in the future. This is currently being addressed (see "Status" above).

Suitability They are suitable for "real", large, Ada software development applications.

External users:
None.

Applications:
Any Ada software development projects.

Availability:
Free from the STARS repository, but must also have the IRIS software for the existing version.

Additional remarks:
None.

B-Toolkit Specification and Design Assistant

Participants (project leader, group members):
Software Engineering Section, Information Science and Engineering Branch, Sunbury Research Centre; J. R. Abrial

Survey contact:
Ib Sørensen,
BP International Ltd.,
Sunbury Research Centre,
Chertsey Road,
Sunbury-on-Thames,
Middx TW16 7LN, U.K.
(tel. 0932 763686; fax 0932 764469)

Brian Weeks,
BP Innovation Centre, Pennel Building,
Sunbury Research Centre,
Chertsey Road,
Sunbury-on-Thames,
Middx TW16 7LN, U.K.
(tel. 0932 763482; fax 0932 763189)

Level of effort (person-years, duration):

Description:
The toolkit supports the development of software, from specification to design and coding, using the B Method. The B Method includes the Abstract Machine Notation, a formal notation belonging, like VDM and Z, to the state-based school of specification. Tools are provided within an integrated window-based environment for:

- editing, syntax and type checking of AMN specifications and refinements

- automatic production of LaTeX documentation of AMN specifications and refinements

- generation of standard AMN specifications from a high level design language

- animation of AMN specifications

- automatic generation of proof obligations for the correctness of AMN specifications and refinements

- carrying out automatic and interactive proofs of the correctness of AMN specifications and refinements

- configuration management of formal developments.

The alpha test version runs on SUN systems under versions 3.x and 4.x of the SunOS operating system as well as on IBM RS6000/AIX and HP/Unix machines.

Accomplishments:

Early versions of elements of the B-Toolkit have been in use over the past few years in software development projects:

- A data base base of a hundred million objects was implemented for the French National Census using the toolkit code generator. The project achieved SUN/IBM portability.

- A speed control system was produced by GEC Alsthom for the French National Railways using an early version of the toolkit.

- A speed control system for the Calcutta underground was developed from specification to implementation by GEC Alsthom using the current version of the toolkit, with code produced by an early version of the toolkit code generator.

- A 45000-line refinery graphics system was developed by BP using the B-Method; 28000 lines of the code were generated automatically.

Published articles or reports:

- M. Lee, P. N. Scharbach and I. H. Sorensen, *Engineering Real Software Using Formal Methods*, Proceedings of the Fourth Refinement Workshop, ed. J. Morris and R. Shaw, Springer-Verlag, 1991

- J. R. Abrial, M. Lee, D. Nielson, P. N. Scharbach and I. H. Sorensen, *The B-Method*, Proceedings of the VDM91 Conference, Springer-Verlag, 1991.

- Alpha-test documentation, BP International Ltd.:

B-Technology Technical Overview

B-Method Overview

B-Method Abstract Machine Notation Summary

B-Toolkit Installation Guide

B-Toolkit User's Manual

Status:
Undergoing external alpha-testing until 31st December 1992.

Start date:

Completion:

Future developments:
The commercial version of the toolkit will incorporate a generator and a translator for the automatic generation of C code.

Strengths, weaknesses and suitability:

External users:
Alpha-test participants (11 external European organisations and 3 U.K. universities).

Applications:
Development of safety-critical, high-integrity and high-quality software systems.

Availability:
Expected to become commercially available in early 1993.

Additional remarks:

B-Tool

Participants (project leader, group members):
J. R. Abrial, BP International Ltd.

Survey contact:
The Distribution Manager,
Edinburgh Portable Compilers Ltd.,
17 Alva Street,
Edinburgh EH2 4PH,
U.K.
tel. 031 225 6262;
fax 031 225 664

Level of effort (person-years, duration):

Description:
The B-Tool is a generic platform to support pattern matching and rule-rewriting for the manipulation and analysis of symbolic text. A number of built-in mechanisms are provided to enable the construction of formal proof systems on the tool. The tool's behaviour is controlled by rule bases

known as 'theories', whose application is guided by "tactics". Specific proof strategies may be prescribed using appropriate combinations of theories and tactics. The tool can also be used, through rule-based programming, to construct and manipulate formal objects other than proofs.

Accomplishments:

Various proof systems have been implemented on the tool, including a natural deduction system and a decision procedure for propositional logic.

The B-Tool is the platform for a number of formal software development tools, including the B-Toolkit and zedB.

Published articles or reports:

- B-Tool User Manual

- B-Tool Tutorial

- B-Tool Reference Manual,

- available from Edinburgh Portable Compilers Ltd.

Status:

Version 1.1 completed in 1991.

Start date:

Completion:

Future developments:

The next release will appear with a Motif user interface.

Strengths, weaknesses and suitability:

External users:

Industrial and Government organisations, Universities.

Applications:

Theorem proving, tool building.

Availability:

Commercially available from Edinburgh Portable Compilers Ltd. The B-Tool runs on Sun systems under versions 3.x and 4.x of the SunOS operating system and on IBM RS6000 workstations.

Additional remarks:

CABERNET An environment for the specification design and validation of real time-systems.

Participants (project leader, group members):

It is being carried out with the scientific guidance of Prof. Carlo Ghezzi and Prof. Mauro Pezze (mpezze@ipmel2.polimi.it) and with the aid of students from Politecnico di Milano and from CEFRIEL.

Survey contact:

Prof. Mauro Pezze (mpezze@ipmel2.polimi.it)

Level of effort (person-years, duration):
   not given

Description:
   CABERNET is a software engineering environment for the specification design and validation of time dependent systems. CABERNET provides a rigorous language in which formal description of functional and time constraints for a system may be specified. The language is based on a high-level Petri net formalism called ER nets [GMMP91]. The environment provides an editor and an animated simulator for the graphical execution of specifications. The environment may be tailored for different application domains by defining the appropriate abstractions using the meta-editing facilities. Moreover, CABERNET is an open environment, since new user defined tools may be programmed using ER nets and added to the basic environment.

Accomplishments:
   The CABERNET project was born on September 1989 and at present time, ver 2.0 has been completed and runs under ULTRIX 4.2 with OSF/Motif. New developments to be added in the environment include modularization mechanisms and simulation of time during execution. These extensions will be completed by September 1993.

Published articles or reports:
   [GMMP91] C. Ghezzi, D. Mandrioli, S. Morasca, M. Pezze, "A Unified High Level Petri Net Formalism for Time Critical Systems". IEEE Transactions on Software Engineering, March 1991.

CADiZ

Participants (project leader, group members):
   Professor John McDermid (Project Leader), Dr Ian Toyn, David Jordan and Brian Sharp

Survey contact:
   David Jordan
   York Software Engineering Limited
   University of York
   YORK
   YO1 5DD
   England

   Tel: +44 904 433741
   Fax: +44 904 433744
   E-mail: yse@minster.york.ac.uk

Level of effort (person-years, duration):
   4 man years over 3.5 years.

Description:
   CADiZ is a UNIX based suite of tools designed to check and typeset Z specifications. CADiZ also supports previewing and mouse-driven interactive investigation of the properties of Z specifications on bit-map workstations.

For example: errors reports can be related to the corresponding parts of schemas; types of variables and signatures of schemas can be inspected; use of declarations can be "tracked"; schema calculus expressions can be expanded.

CADiZ supports the standard Z core language and the mathematical toolkit. In addition CADiZ provides support for user defined operators, documents (which allow multi part, structured, specifications to be handled), versions, fancy symbols and the automatic production of a specification index.

Typesetting support is provided through both troff and TEX.

Accomplishments:
CADiZ is used throughout the world to support teaching, research and development activities. Of particular note, CADiZ forms part of the wide-spectrum Z toolset being developed under the ZIP Project.

Published articles or reports:

- D.T.Jordan, J.A.McDermid and I.Toyn, *CADiZ Computer Aided Design in Z*, in Z User Workshop, Oxford 1990, pp93-104, Springer-Verlag

- I.Toyn, *CADiZ Quick Reference Guide*, York Software Engineering Limited, 1991

Status:
First commercial release March 1991. Ongoing research, development and application.

Start date:
October 1988.

Completion:
Ongoing.

Future developments:
An X Window System-based version of CADiZ has just been completed which allows CADiZ to be made available on a wider range of platforms (see Availability).

Work on TEX typesetting support is about to go to beta-test. This version of the tool will be fuzz input compatible.

At the University of York, where CADiZ was originally developed, work is currently being done to provide proof support within the CADiZ environment. Preliminary results have been encouraging and a prototype tool has been produced this was demonstrated at the 6th Annual Z User Meeting held in York in December 1991. Development will continue throughout 1992.

In conjunction with the proof support development, CADiZ is being restructured with the aim of providing an open Z environment into which it will be easy to integrate other (third-party) tools.

Strengths, weaknesses and suitability:
CADiZ has been found to be very easy to use by all types of users, from Z novices to experienced specifiers. The previewing and interactive investigation facilities are unique and particularly powerful feature.

Lack of proof support is an obvious weakness. However, as noted above, this is currently being addressed.

External users:
CADiZ is in day-to-day use at both academic and industrial sites in the UK, the USA, France, Italy and Australia.

Applications:
Various.

Availability:
CADiZ is available on Sun3 and Sun4 platforms. Enquiries about potential ports are welcome and should be directed to David Jordan.

Additional remarks:
York Software Engineering won a DTI SMART (Small firms Merit Award for Research and Technology) Award in 1991 to study tools to support the refinement of Z specifications to Ada.

OYSTER–CLAM

Participants (project leader, group members):
Alan Bundy, Alan Smaill, Ian Green, Jane Hesketh, Andrew Ireland, Helen Lowe and Gordon Reid.

Previous participants include David Basin, Frank van Harmelen, Christian Horn, Peter Madden, Andrew Stevens and Lincoln Wallen.

Survey contact:
Gordon Reid, Department of Artificial Intelligence,
University of Edinburgh, 80 South Bridge,
Edinburgh, EH1 1HN, Scotland.
Tel: (031) 650 2728
email: gordon@uk.ac.ed.aisb
or
Andrew Ireland,
University of Edinburgh (same address as above),
Tel: (031) 650 2721
email: air@uk.ac.ed.aisb

Level of effort (person-years, duration):
Approximately six person years (since 1988)

Description:
OYSTER[2] is an interactive proof editor closely based on the Cornell NuPRL system but implemented in Prolog. The object-level logic is a version of Martin-Löf Type Theory (a constructive higher-order logic including induction) in a sequent-calculus formulation. Proofs are constructed in a top-down fashion by application of the rules of inference. Tactical proof, notational definitions and libraries of theorems are supported.

CLAM[3] is a planning system built on top of OYSTER to turn the interactive proof editor into a fully automatic theorem proving system. CLAM uses specifications of OYSTER tactics in the search for a proof. These specifications are called *methods*. A method acts as a heuristic operator which can capture the essential preconditions of a tactic. The search for a proof generates a tree of methods which we call the *proof plan*. The process of composing methods is called *proof planning*. Plan execution corresponds to using the tactic component of a proof plan to control the application of inference rules in OYSTER.

Accomplishments:

The OYSTER-CLAM system has been applied to the domain of inductive proofs [1]. The main source of examples has been the Boyer & Moore corpus of theorems. The system can however automatically prove synthesis theorems such as prime factorisation which represents an improvement on the Boyer & Moore theorem prover.

Published articles or reports:

1 Bundy, A. and van Harmelen, F. and Hesketh, J. and Smaill, A. *Experiments with Proof Plans for Induction*, Journal of Automated Reasoning, 7:303-324, 1991.

2 Horn, C. *The Nurprl Proof Development System.* Working paper 214, Dept. of Artificial Intelligence, Edinburgh, 1988. The Edinburgh version of Nurprl has been renamed Oyster.

3 van Harmelen, F. *The CLAM Proof Planner, User Manual and Programmer Manual*, Technical Paper TP-4, Dept. of Artificial Intelligence, Edinburgh, 1989.

Status:

A distribution version is available, next release expected early summer 1992.

Start date:

1988

Completion:

Ongoing

Future developments:

Areas of possible development for the future include:

- Cooperative theorem proving

- An Intelligent Tutoring System for inductive proof

- Using proof plans to build flexible decision procedures

- Automatic proof plan patching

- A generic proof planning framework

Strengths, weaknesses and suitability:

The OYSTER-CLAM system supports totally automated proof of simple inductive theorems. The system is weak on non-inductive proofs and theorems involving existential quantification. Only a very simple interactive interface is provided.

External users:
Andrew Stevens PRG Oxford, David Basin Max-Planck Institute Saar-bruecken, Richard Boulton HOL Group Cambridge.

Applications:
Inductive theorem proving.

Availability:
Contact Gordon Reid (address as shown above).

Additional remarks:
This research is sponsored by SERC grant GR/F/71799.

Clio

Participants (project leader, group members):
Mandayam Srivas and Mark Bickford

Survey contact:
M.K. Srivas
ORA Corporation
301A Dates Drive
Ithaca, NY USA 14850-1313.

Level of effort (person-years, duration):
4 staff years

Description:
Clio is a system for proving properties about programs written in an executable functional language, Caliban. Caliban is a higher order, polymorphic, lazy functional language similar to Miranda (Miranda is a trademark of Research Software Limited). A property to be proved is expressed in the Clio assertion language as an arbitrary first order predicate calculus formula built from atomic literals. An atomic literal is an equation on Caliban expressions.

The basic proof technique of the Clio prover is normalization, i.e., simplifying expressions on the two sides of an equation to a common form to prove the equation. The prover supports a set of proof tactics that are useful in conjunction with normalization to prove more complex formulae. Some of the proof tactics available are: case analysis, structural induction, fixpoint induction, and proof by contradiction. The user can use the prover in interactive or automatic mode. In the interactive mode, Clio prompts the user with the available choices in proof tactics, and the user must make the appropriate selection until Clio proves the formula or discovers a contradiction. In the automatic mode, it makes the selection on its own based on a built-in strategy.

Accomplishments:
An interactive verification system for a modern functional language has been built. Recently, several features have been added to make proof construction automatic in many situations.

Published articles or reports:

- M. Bickford, C. Mills, and E.A. Schneider, *Clio: An Applicative Language Based Verification System.* Technical Report TR 89-13, ORA Corporation, September, 1989, 301A Harris B. Dates Drive, Ithaca, NY 14850.

- Mandayam Srivas and Mark Bickford, *Formal Verification of a Pipelined Microprocessor.* IEEE Software, September, 1990.

Status:

The theorem prover implementation is stable and has been and is being used on several case studies, primarily in the area of hardware verification at ORA.

Start date:

July 1985.

Completion:

July 1989.

Future developments:

We plan to implement a graphical user-interface that supports a friendlier and more efficient interaction with the user.

Strengths, weaknesses and suitability:

Strengths Provides expressive power of a modern non-strict functional language for specification. Can specify reason about partial functions and streams. Support for efficient rewriting and induction principles.

Weaknesses Does not have good support for building proof tactics. Does not yet have a rich theory library.

External users:

Clio has been used by several PhD students at Syracuse University for their dissertation work in hardware verification.

Applications:

Clio, along with a support tool Spectool (also developed at ORA), has been used on three large hardware design verification case studies. The first case study involved the verification of a pipelined microprocessor (MiniCayuga) design. The second involved the verification of fault tolerant property of a fault tolerant system involving four extended versions of MiniCayuga. It is currently being used to verify a key hardware component of a fault tolerant parallel computer being developed at Charles Stark Draper Laboratory.

Availability:

Clio is implemented in C and runs under UNIX based system. It is available for distribution with permission of our contractors.

Additional remarks:

ConceptBase

Participants (project leader, group members):
Matthias Jarke, Manfred Jeusfeld, Thomas Rose (many others)

Survey contact:
Matthias Jarke
Informatik V, RWTH Aachen
Ahornstr. 55
5100 Aachen, Germany
email: jarke@informatik.rwth-aachen.de

Level of effort (person-years, duration):
approximately 10 person years (since 1986)

Description:
A deductive object base management system based on the knowledge representation language Telos which integrates object-oriented structure, deductive rules and integrity constraints from deductive databases, and an interval-based time calculus. This was used as a basis for defining formal process models through which heterogeneous development tools can be controlled and integrated. ConceptBase also contains various optimization techniques for query processing and integrity checking on design process knowledge bases so that it can be used for rapid prototyping of information systems. A hypertext-oriented usage environment with textual, structure-sensitive, and graphical interface tools is provided as well as specialized tools for version and configuration management.

Accomplishments:
ConceptBase has been applied for the integration of the DAIDA information systems development and maintenance environment, and for the development of a process model for a hypertext co-authoring environment called CoAUTHOR. Both involved a specification of these environments in the system. Furthermore, a complete model of development-in-the-large at a conceptual level has been developed and evaluated in the system.

Published articles or reports:

- J.Mylopoulos et al.: Telos *A language for representing knowledge about information systems.* ACM Trans. Information Systems 8, 4, 1990.

- M. Jarke et al.: DAIDA *An environment for evolving information systems.* ACM Trans. Information Systems 10, 1, 1992

- U. Hahn et al.: *Teamwork support in a knowledge-based information systems environment.* IEEE Trans. Software Eng. 17, 5, 1991

- T. Rose et al.: *A decision-based configuration process environment.* IEE Software Engineering Journal 6, 5, 1991.

- M. Jarke (ed.): *ConceptBase V3.0 User Manual, Technical Report,* University of Passau, Germany, 1991.

Status:

Research prototype, distributed to about 20 universities and industry research labs for application experiments

Start date:
Fall 1986

Completion:
First version Nov. 1987, currently working on version 4

Future developments:
Integration of non-monotonic reasoning facilities and hypermedia for informal specifications

Strengths, weaknesses and suitability:

Strengths Suited for metadata management and integration, emphasis on process support including teamwork

Weaknesses Still limited knowledge base size, limited to deductive database like reasoning procedures, no sophisticated theorem-proving

External users:
IRIS Network of Excellence in Canada (repository/reusability research) New York University (requirements propagation in teamwork setting) various universities for teaching industrial and government labs for requirements analysis support for Total Quality Management in production engineering

Applications:

Availability:
For research purposes on copy-fee basis, needs some licensed base software

Additional remarks:
none given

DECspec

Participants (project leader, group members):
Joe Wild, Gary Feldman, Bill McKeeman (DEC Software Development Technologies), Jim Horning, Kevin Jones (DEC Systems Research Center), John Guttag, Steve Garland (MIT), Jeannette Wing (CMU)

Survey contact:
Joe Wild
Digital Equipment Corporation ZKO2-3/N30
110 Spit Brook Road
NASHUA, NH 03062
U.S.A.

Level of effort (person-years, duration):
3-4 person-years to date

Description:
Syntax and type checkers for the Larch Shared Language (LSL 2.3), and a Larch/C interface language (LCL 1.0)

Accomplishments:
Tools now available for research and education. Familiarity with the Larch methodology transferred from research into product organization. Technical challenges from product group transferred back to research.

Published articles or reports:

- Book in preparation
- J.V. Guttag, J.J. Horning, and A. Modet, *Report on the Larch Shared Language: Version 2.3* DEC/SRC Report 58, 1990
- J.V. Guttag and J.J. Horning *Introduction to LCL, A Larch/C Interface Language* DEC/SRC Report 74, 1991

Status:
Maintenance only

Start date:
July 1990

Completion:
Ramped down after June 1991

Future developments:
Open

Strengths, weaknesses and suitability:
Suitable for experimentation with the Larch methodology and specifications for interfaces in (or called from) C programs. User interface (error reporting) still very rough. Not very many miles on the tires.

External users:
FTPed by about 100 sites; no data on actual usage

Applications:
Needed

Availability:
Anonymous FTP from gatekeeper.dec.com; no fee; liberal license for research and educational use

Additional remarks:
We are eager for feedback

DynaMan (Dynamic modelling based Management tool)

Participants (project leader, group members):
Alfonso Fuggetta, Luigi Lavazza

Survey contact:
CEFRIEL,
via Emanueli 15,
20126 Milano,
Italy tel.
+39-2-6610-0083
e-mail: alfonso@mailer.cefriel.it, lavazza@mailer.cefriel.it

Level of effort (person-years, duration):

Description:

A few years ago, Abdel-Hamid and Madnick suggested the use of system dynamics as a rigorous, innovative and effective technique to model the software development processes. They showed that system dynamics can be used not only to achieve better cost estimations, but also as a general way to evaluate different managerial strategies.

DynaMan is a tool designed to support the software manager in the usage of dynamic models: it supports the creation, customization, maintenance and simulation of dynamic models of the software development process. A schematic representation of the architecture of DynaMan is given in [2]. The present version of DynaMan provides:

- a syntax directed editor to input equations of a dynamic system and to incrementally build the internal representation of the model. Some semantic checks are also performed;
- a math kernel, whose task is the execution of dynamic systems: it evaluates the equations constituting the model using the information stored in the syntax tree and in the symbol table. Execution is performed by computing (step by step) the values of variables in time;
- a graphic display subsystem that provides graphic visualization of simulation results.

The development of DynaMan started in November 1990; three man-years were needed to complete version 1.1. The tool is developed under Ultrix using C++ and DecWindows. Development is performed at CEFRIEL, an educational and research institute located in Milan and supported by a consortium of local industries and universities. The results of the research activities are freely available to the participants in the consortium.

The basic functionality currently offered by DynaMan is provided by other commercial products. The present version of DynaMan is being used as a starting point for further experimentation and to test the applicability of techniques derived from the automatic control area. In particular we are trying to define parametric identification techniques that can be applied to highly non-linear systems like the model proposed by Abdel-Hamid and Madnick. It is necessary to have some parametric identification method in order to properly tune models: even though a model is structurally correct, it may simulate poorly the real system if the parameters' values are not correct. If our research of identification techniques will be successful, we will include support for these techniques in DynaMan.

Accomplishments:

Published articles or reports:

1 A. Fuggetta, L. Lavazza *Research directions in quantitative software process modelling: the system dynamics approach* Proc. AICA national congress 1991 and CEFRIEL Report RT91001 February 1991.

2 M. Bonamici, A. Laria, *A Dynamic System based Tool for Software Process Modelling*, CEFRIEL Report RI91076, June 91

3 M. Bonamici, A. Laria, *DynaMan v.1.0 User Manual* CEFRIEL Report RI91093, July 1991.

4 A. Barbieri, A. Fuggetta, L. Lavazza, M. Tagliavini, *DynaMan: A Tool to Improve Software Process Management through Dynamic Simulation*, to be published in Proc. CASE 92 and CEFRIEL Report RT91066 December 1991

The ESSE (Environment Supporting Schema Evolution) project

Participants (project leader, group members):
Alberto Coen-Porisini, Luigi Lavazza, Roberto Zicari

Survey contact:
Politecnico di Milano
Dipartimento di Elettronica
Piazza Leonardo da Vinci 32,
20133 Milano,
Italy
tel.+39-2-2399-3634
e-mail: coen@ipmel2.elet.polimi.it

CEFRIEL, via Emanueli 15,
20126 Milano,
Italy
tel. +39-2-6610-0083
e-mail: lavazza@mailer.cefriel.it

J. W. Goethe-Universitaet
Fachbereich Informatik (20)
Robert-Mayer-Str. 11-15,
D-6000 Frankfurt am Main 11,
Germany
tel. +49-69-798-8212
e-mail: zicari@informatik.uni-frankfurt.de

Level of effort (person-years, duration):

Description:
The aim of the ESSE project is to provide an environment supporting database design activities, such as logic design, schema modification, schema correctness checking, etc. The project started in June 1991; as a first step we faced the problem of ensuring the behavioral correctness of a schema. We consider systems (such as the O2 database system and the Eiffel programming environment) providing redefinition of a method based on the covariance subtype rule and allowing the redefinition of the domain of the method: in general these systems do not provide run-time type safety in the presence of late binding. It is well known that for these systems deciding whether a schema is type safe is an undecidable problem; however we have demonstrated in [2] how schemas can be statically type checked by defining a data flow technique which detects all possible type errors and gives sufficient conditions for ensuring type safeness. In [3] we presented in detail the

algorithms which make use of such data flow technique. The first phase of our project consists in the realization of a prototype tool, whose architecture is described in [4], that will be able to perform the static type checking of O2 schemas in order to verify their behavioral correctness. This tool will complement the functionalities of the compiler of the schema definition language of O2. An analogy can be made with the lint software utility which performs additional check not provided by a standard C compiler. Although we chose O2 as the target system of the prototype, only the schema interpreter (realized using lex and yacc) is target dependent, so it will be enough to modify such subsystem to adapt the tool to other systems or languages. The tool is developed under Ultrix using C++. Development is performed at CEFRIEL; it began in November 1991 and will require a two man-years effort; the first release should be available in October 1992. Since CEFRIEL is a non-profit organization the results of the research activities are freely available to the participants in the consortium. The future evolution of the project aims at the development of an advanced integrated software environment with main goal to support, help, and advise the data engineer in performing schema modifications for an object-oriented database system. In this sense ESSE will be a step towards the development of tools for the logical design of an object-oriented database.

Accomplishments:

Published articles or reports:

1. Coen-Porisini A., Lavazza L., Zicari R. *Updating the schema of an object oriented database* IEEE Data Engineering Bulletin, June 1991, Vol. 14, No. 2

2. Coen-Porisini A., Lavazza L., Zicari R. *Verifying Behavioral Consistency of an Object-Oriented Database Schema*, Politecnico di Milano Research Report No. 91-054, November 1991

3. Coen-Porisini A., Lavazza L., Zicari R. *Static Type Checking of Object-Oriented Databases*, Politecnico di Milano Research Report No. 91-060, November 1991

4. Coen-Porisini A., Lavazza L., Zicari R. The ESSE project: an overview, to appear in the Proc. of The Second Far-East Workshop on Future Database Systems, April 27-28, 1992, Heian Shrine, Kyoto, Japan

ESTIM (Estelle SimulaTor based on an Interpretative Machine)

Participants (project leader, group members):
Pierre de Saqui-Sannes, Jean-Pierre Courtiat, Rosvelter J. Coelho da Costa, L.F. Rust da Costa Carmo (plus many others in the past, notably Vitorio B. Mazzola).

Survey contact:
Pierre de Saqui-Sannes
LAAS-CNRS
7 avenue du Colonel Roche

31077 Toulouse cedex
France
Tel: +33 61.33.62.44
Fax: +33 61.33.64.11
E-mail: pdss@laas.fr

Level of effort (person-years, duration):
Design: 1.5 men/year ;
Users: 2 men/year (since 1985)

Description:
ESTIM is a validation environment for protocols and distributed systems specified using Estelle*. Estelle* adds a rendezvous mechanisms on the top of Estelle [IS 9074], a message-passing Formal Description Technique based upon Extended Finite State Machines and Pascal.

ESTIM basically relies on an Estelle* interpreter. It offers facilities for interactive simulation, including the display of Estelle objects, several transition firing policies as well as an estimation of the simulation coverage.

The main potential of ESTIM lies in the way it exploits the reachability graph of an Estelle* specification as a labelled transition system. Indeed, the latter is reduced using an equivalence relation, e.g. trace, observational or test equivalence. Thus, from a protocol layer specification written in Estelle*, one becomes able to automatically derive an abstract view characterizing the service provided by that protocol.

Accomplishments:
Validation of various real-size protocols including Transport T70, a subset of XTP, MMS (Manufacturing Message Services), the FIP fieldbus protocol as well as the driving system of a Flexible Assembly Cell. The latter experience was concluded generating C-code from the Estelle source, using the Estelle WorkStation developed within ESPRIT-Sedos/Estelle/Demonstrator project.

Published articles or reports:

- J.P. Courtiat, P. de Saqui-Sannes, *ESTIM: an Integrated Environment for the Simulation and Verification of OSI protocols specified in Estelle,* invited paper to appear in the Special Issue of the Journal of Computer Networks and ISDN systems on Tools for Protocol Engineering, 1992.

- P. de Saqui-Sannes, *The ESTIM User-Manual,* LAAS internal report, 1990.

Status:
Public release accessible via anonymous ftp.

Start date:
September 1985

Completion:
Deliverable of ESPRIT-SEDOS project: November 1988.
Public release: October 1990.

Future developments:

Internal reorganization and partial compilation for increased running-time performances. Extensions to test sequence generation.

Strengths, weaknesses and suitability:

Strengths easily accessible for newcomers to Estelle and protocols.

Weakness memory consumption due to interpretation in ML.

Suitability Courses on distributed computing and computer networks. Prototyping of new distributed algorithms and protocols.

External users:
Electricite de France, Legrand, Universite Paul Sabatier (Toulouse), University of Clermont-Ferrand, Federal University of Lausanne, University of Montreal, University of Berkeley, UFSC (Brasil), University of Sao Paulo, University of Rio de Janeiro.

Applications:
ftp laas.laas.fr
login: anonymous
passwd: your identifier
cd /pub/estim
get README

Availability:
Partly supported by the ESPRIT-SEDOS project (ST 415).

Additional remarks:

EVES

Participants (project leader, group members):
Dan Craigen, Sentot Kromodimoeljo, Bill Pase, Irwin Meisels, Mark Saaltink

Survey contact:
Dan Craigen
Odyssey Research Associates
265 Carling Ave., Suite 506
Ottawa, Ontario K1S 2E1
CANADA

Phone: (613) 238-7900
email: dan@ora.on.ca

Level of effort (person-years, duration):
15 person years (approximately)

Description:
Eves is a verification system based on first-order, untyped set theory (with the Axiom of Choice). Programs are specified, implemented and proven using the language Verdi. A denotational description of the linguistic components for specifying and implementing programs has been developed. A

formal characterization of the proof obligations resulting from each declaration, so as to maintain a semantic conservative extension property, has also been developed. (This includes a mathematical justification of the analysis performed by the Verdi verification condition generator.) Amongst other constructs, Verdi supports a library facility, mutual recursion, a form of strong typing for executable constructs and various constructs to support the expression of general mathematical concepts.

The theorem prover NEVER, while strongly influenced by its predecessor m-NEVER, has been modified to mirror the change in logical framework from m-EVES to Eves. Eves is implemented in Common Lisp and has been successfully run on Symbolics, Sun 3, Sparcstation, Data General Aviion, Macintosh IIx and on a Vax.

Accomplishments:
Eves is based on a solid mathematical foundation and has a powerful theorem proving tool. To date, the largest Eves application has been used to specify and implement significant portions of an interpreter for PICO. The resulting specification, implementation, and system commands consists of over 11,000 lines.

Published articles or reports:

- Dan Craigen, Sentot Kromodimeoljo, Irwin Meisels, Bill Pase, Mark Saaltink. *Eves: An Overview* VDM'91, Lecture Notes in Computer Science, # 551, Springer-Verlag.

- Mark Saaltink. *Z and Eves* Z User Workshop, York 1991 Workshops in Computing, Springer-Verlag.

- Dan Craigen. *The Verdi Reference Manual.* TR-91-5429-09a, Odyssey Research Associates, September 1991.

- Mark Saaltink. *The Mathematics of Verdi.* TR-90-5429-10a, Odyssey Research Associates, November 1990.

- Mark Saaltink. *Alternative Semantics for Verdi.* TR-90-5446-02, Odyssey Research Associates, November 1990.

- Sentot Kromodimoeljo, Bill Pase. *Using the Eves Library Facility: A PICO Interpreter.* TR-90-5444-02, Odyssey Research Associates, February 1990.

Status:
Version 1.0 delivered March 1990. Ongoing research, development and application.

Start date:
November 1987

Completion:
Ongoing

Future developments:
Continued evolution of the system's capabilities and investigation of its inclusion into the broader system's engineering perspective.

Strengths, weaknesses and suitability:
Sound mathematical basis, expressive language, and a powerful theorem prover. On the other hand, Eves is an isolated tool and Verdi is not a mainstream development language.

External users:
A few, but still in the early stages of usage.

Applications:
PICO Interpreter; Z toolkit incorporated; General framework for mathematical modelling.

Availability:
Available both for research use and commercial application.

Additional remarks:

Formal Development Methodology (FDM)

Participants (project leader, group members):
Deborah Cooper, Steve Eckmann, Paul Eggert, Jonathan Gingerich, John Scheid, Dewey V. Schorre, Traci Wheeler (plus several others in the past)

Survey contact:
Deborah Cooper,
Dir. Advanced Technology,
Paramax Systems Corporation, 12010 Sunrise Valley Drive,
Reston, VA 22091
USA
Tel: (703) 620-7778
email: cooper@rtc.reston.unisys.com

Level of effort (person-years, duration):

Description:
FDM is a tool suite for formally specifying and verifying software systems. Ina Jo, the FDM specification language, is a typed extension of first order logic plus set theory. Ina Flo, the FDM information flow tool, partially automates covert channel analysis of Ina Jo specifications. Ina Go, the FDM Model Executor, provides an environment for executing Ina Jo specifications. The Interactive Theorem Prover (ITP) is an interactive proof checker for formally verifying Ina Jo specifications. The ITP has been replaced by Nate, a Natural deduction Automated Theorem proving environment; Nate completes many proofs without human intervention. The underlying logic of Nate is based on a well known logic system; soundness and completeness proofs are available. Full documentation for the FDM Tool Suite is available.

Accomplishments:

- FDM is one of two formal verification systems endorsed by the National Computer Security Center for use in achieving A-1 certification under the DoD Trusted Computer System Evaluation Criteria. FDM has been used extensively on many significant Multilevel Secure (MLS) systems, including two systems which have been certified A-1 and two additional systems which are currently undergoing evaluation for certification at class A-1.

- Autodin II, a large secure distributed network.

- The Secure Transaction Processing Executive (STPE), a security kernel for an existing operating system.

- The Job Stream Separator (JSS), a color-change controller for doing periods processing on a large scientific computer system.

- The Kernelized Virtual Machine (KVM), a large MLS operating system based on the IBM VM/370.

- The Computer Operating System/Network Front-End (COS/NFE), a MLS operating system (a kernel and four trusted processes) acting as a secure front-end to an existing network.

- The Secure Release Terminal (SRT) Project, a small security guard for reviewing and sanitizing classified data before release.

- The Paramax Blacker system, a very large MLS distributed network with end-to-end encryption. Blacker has been certified A-1 under the DoD Trusted Computer System Evaluation Criteria.

- The Boeing MLS Local Area Network, which provides multilevel connectivity between terminals, workstations, and hosts over an IEEE 802.3 trunk, as well as multilevel circuit-switch control. The access control device (the Secure Network Server) has completed evaluation as an A1 network component under the NCSC Trusted Network Interpretation. A second generation of Secure Network Servers, providing embedded network management functionality and interconnectivity between networks is currently in the Design Analysis Phase of evaluation by the National Computer Security Center.

- The Gemini Trusted Network Processor (GNTP), a MLS communication processor platform with a high-performance multi-processor executive. The GNTP is currently under evaluation as a class A1 network component. FDM was also used in the formal modeling and specification of the Gemini Multiprocessing Secure OS (GEMSOS) Trusted Computing Base (TCB), a MLS operating system.

Published articles or reports:

- *An Information Flow Model for FDM*, Steven Eckmann, Paramax Systems Corporation, TM 8416/002, March 1992.

- *A Retrospective on the VAX VMM Security Kernel*, Paul Karger et al., IEEE Transactions on Software Engineering, vol. 17, 11, November 1991, pages 1147-1165.

- *Introduction to the Gemini Trusted Network Processor*, Proceedings of the NIST/NCSC 13th Computer Security Conference, Baltimore, Md. October 1990, pages 211-217.

- *Industrial Experience Using FDM*, Deborah Cooper, ICSE12 Workshop on Industrial Experience Using Formal Methods, March 1990.

- *Extending Ina Jo with Temporal Logic*, Jeannette Wing and Mark Nixon, IEEE Transactions on Software Engineering, vol. 15,2, February 1989, pages 181-197.

- *KMN: Natural Deduction for An Automated Theorem Proving System*, Robert Martin, Paramax Systems Corporation, TM 8477/001, January 1989.

- *FDM User Guide*, Paul Eggert et al., Paramax Systems Corporation, TM 8486/000, December 1988.

- *Towards a Formal Basis for the Formal Development Method and the Ina Jo Specification Language*, Daniel Berry, IEEE Transactions on Software Engineering, vol. SE-13,12, February 1987, pages 184-201.

- *Shared Resource Methodology: An Approach to Identifying Storage and Timing Channels*, Richard Kemmerer, ACM Transactions on Computer Systems, vol. 1, 3, August 1983, pages 256-277.

Status:
Maintenance and enhancements.

Start date:
1974

Completion:
Transition from research to production use, approx. 1983-85; enhancements continued.

Future developments:
not stated

Strengths, weaknesses and suitability:

Strengths FDM has been successfully used on nearly a dozen significant MLS system development efforts, notably by system designers and engineers whose primary training was not in formal methods. FDM has been reported to be very useful in identifying and locating design flaws during specification.

Weaknesses FDM does not provide support for modularity and composition, or for specifying and verifying temporal properties.

External users:
FDM has been distributed to more than 60 user sites and has been used for several large development projects at Boeing Aerospace Corporation, Gemini Corporation, Digital Equipment Corporation, and Unisys Corporation, among others.

Applications:

Availability:
FDM tools are distributed without cost to users; some restrictions apply.

Additional remarks:

Gypsy Information Flow Tool (GIFT)

Participants (project leader, group members):
John McHugh, Robert L. Akers, Bret Hartman, Tad Taylor, Art Flatau,
Craig Singer

Survey contact:
Robert L. Akers
Computational Logic Inc.
1717 W. Sixth St., Suite 290
Austin, TX 78703
phone: (512) 322-9951
email: akers@cli.com
fax: (512) 322-0656

Level of effort (person-years, duration):
Roughly two person years over roughly two years

Description:
Covert channels are mechanisms that allow users of a supposedly secure
computer system to communicate with each other in violation of the system's
security policy. The Gypsy Information Flow Tool (GIFT) embodies one
method for identifying and evaluating such channels. GIFT is based on
an analysis of the specification of the trusted computing base (TCB) of a
secure system. The tool also provides the analyst with a detailed description
of the information dependencies that are present within the TCB and the
circumstances under which the potential channels identified by the GIFT can
be exercised. The GIFT is designed to analyze TCB specifications written
in the Gypsy language and is a fully integrated component of the Gypsy
Verification Environment.

The Gypsy Information Flow Tool is based on a combination of a dependency
analyzer and a security formula generator. Informally, we define dependency
analysis to be the process of tracing the flow of information into, through,
and out of a system based upon a formal specification for that system. In
the context of the GIFT, we provide a more formal definition of dependency
analysis that relies upon Gypsy language semantics and Gypsy specification
conventions designed to model state machines and security policies. The
information provided by dependency analysis is combined with user supplied
information about the security levels of system entities to generate formulas
which, if proven to be true, ensure that the information flows identified
during the dependency analysis conform to the system's security policy.

Accomplishments:
The tool has been used to perform flow analysis tasks on TRW's Army
Secure Operating System (ASOS) and on Honeywell's Logical Co-Processor
Kernel (LOCK).

The Honeywell LOCK project is using non-interference proofs to help establish the security of their system. As with any non-interference proof,
it is important that the concept of what is "visible" to a user be captured
correctly, since proofs are performed with respect to that definition. The
dependency analysis capabilities of the GIFT were used to help validate the

completeness of the visibility definition.

Published articles or reports:

- B.A. Hartman. *Non-Interference and Information Flow Analysis: A Hybrid Approach.* In Proceedings of the Computer Security Foundations Workshop, Franconia, NH, June 1989.
- J. McHugh and R.L. Akers. *Specification and Rationale for the Implementation of an Analyzer for Dependencies in Gypsy Specifications.* Technical Report N° 15a, Computational Logic, Inc., Austin, Texas, November 1987.
- J. McHugh, R.L. Akers. *A Formal Justification for the Gypsy Information Flow Tool.* Technical Report 13a, Computational Logic, Inc., Austin, Texas, May 1988.
- J. McHugh, R.L. Akers, M.C. Taylor. *GVE User's Manual: The Gypsy Information Flow Tool, a Covert Channel Analysis Tool.* Technical Report 12d, Computational Logic, Inc., Austin, Texas, July 1988.
- O.S. Saydjari, J.M. Beckman, J.R. Leaman. *LOCKing Computers Securely.* In Proceedings of the 10th National Computer Security Conference. Baltimore, MD, 1987.

Status:

The GIFT is a stable component of the Gypsy Verification Environment. No further development is planned.

Start date:

January, 1987.

Completion:

July, 1989.

Future developments:

Although the basis exists for extending its capability into the area of tranquillity analysis, there are no plans for this extension.

The GIFT has its own methods for tracking the effects of the incremental development of the TCB specifications. Since the completion of the GIFT, a new incremental development tool has been installed in the GVE which could supplant the GIFT methods, however the new incremental development tool was not integrated into the GIFT. This could easily be done, but is not on the current task schedule.

Strengths, weaknesses and suitability:

The GIFT represents several advances in the state of the art and practice of flow analysis:

The GIFT theory provided a more satisfactory approach to information flow into structured objects than had been previously utilized. It also extended the conventional theory into the area of dynamic data objects by treating Gypsy sequences, sets, and mappings.

The pragmatic issues of "labelling" objects with their security levels were addressed in a manner much more conducive to formal and automatic analysis than in any previous tool or theory.

The expressive power of the Gypsy language provided a basis for explicit dependency analysis of some cases which were outside the limits of previous tools.

The GIFT was fully integrated into a popular verification system, the GVE. The theory embodied in the GIFT was totally consistent with the theory and semantics of verification in Gypsy, and the software tools were smoothly integrated with the existing system, thus eliminating on both fronts a cognitive dissonance which had plagued other flow analysis tools.

As previously mentioned, the tool does not yet address tranquillity analysis.

External users:
TRW, Honeywell

Applications:
See the Accomplishments section.

Availability:
Distribution of the GIFT requires approval from the USA National Computer Security Center. Direct requests for the GIFT (and the prerequisite government approval) to

Laura Tice
Computational Logic, Inc.
1717 W. 6th St., Suite 290
Austin, Texas 78703
phone: (512) 322-9951
email: tice@cli.com
fax: (512) 322-0656

Additional remarks:
The GIFT tool was supported by the TRW Defense Systems Group in conjunction with work on the Army Secure Operating System (ASOS), Contract N° DAAB07-86-CA032, U.S. Army Communications-Electronics Command. The views and conclusions contained in this document are those of the author(s) and should not be interpreted as representing the official policies, either expressed or implied, of Computational Logic, Inc., TRW Defense Systems Group, or the U.S. Government.

Gypsy Verification Environment (GVE)

Participants (project leader, group members):
Don Good and his associates at The Institute for Computing Science (The University of Texas at Austin) and at Computational Logic, Inc.

Survey contact:
Larry Smith
Computational Logic Inc.
1717 W. 6th St., Suite 290
Austin, TX 78703
phone: (512) 322-9951

email: lsmith@cli.com

fax: (512) 322-0656

Level of effort (person-years, duration):
50-100 Person-Years

Description:

The Gypsy Verification Environment (GVE) is a large computer program that supports the development of software systems and formal, mathematical proofs about their behavior. The environment provides conventional development tools, such as a parser for the Gypsy language. These are used to evolve a library of components that define both the software and precise specifications about its desired behavior. The environment also has a verification condition generator that automatically transforms a software component and its specification into logical formulas which are sufficient to prove that the component always runs according to specification. Facilities for constructing formal, mechanical proofs of these formulas also are provided. Many of these proofs are completed automatically without human intervention. The GVE also contains a Feiertag-style Information Flow Tool.

The Gypsy language is a combined specification and programming language descended from Pascal. It supports a large variety of data types, data abstraction, exception handling, and concurrency. All parts of the language have associated proof rules.

Accomplishments:

The GVE is based on the Floyd-Hoare assertion method for proving properties of computer programs. The Gypsy project was the first application of this method to a reasonably large and usable, general-purpose programming language. The ideas of Floyd and Hoare were extended to cover features such as concurrency.

The GVE was a study in the development of a unified system for computer program development and maintenance. It contains a Gypsy parser, a verification condition generator, an algebraic simplifier, an interactive proof checker, an optimizer (which generates conditions for proving the validity of potential code optimizations by a compiler), a pretty printer, a database manager, an incremental development manager, and an information flow tool. It has contained compilers, translators to other languages (ADA, Bliss), a Gypsy interpreter, and an editor.

The GVE is one of two tools endorsed by the U.S.A. National Computer Security Center (NCSC) for use in achieving A-1 certification under the U.S.A. Department of Defense Trusted Computer System Evaluation Criteria.

Published articles or reports:

- R.L. Akers, B.A. Hartman, L.M. Smith, M.C. Taylor, W.D. Young. *Gypsy Verification Environment User's Manual.* Technical Report 61, Computational Logic, Inc., Austin, Texas, September 1990.

- D.I. Good, R.L. Akers, L.M. Smith. *Report on Gypsy 2.05.* Technical Report 1c, Computational Logic, Inc., Austin, Texas, August 1989.

- And many others, available from Computational Logic, Inc.

Status:
Under maintenance

Start date:
1974

Completion:
Approximately 1985-89.

Future developments:
Starting in 1989, the GVE has been moving from an enhancement phase to a maintenance phase. A new generation of tools to replace it will be developed in the 1990's.

Strengths, weaknesses and suitability:

Strengths Demonstrated effectiveness on sizable projects. Applicable to a wide range of problems, including design and code level proofs.

Weaknesses Because of its research roots, the GVE was never designed to handle the large applications to which it is now being applied. Its implementation shows the signs of many graduate-student programmers over the years. Some components of the GVE, e.g., the proof checker, are based on old (1974) technology.

External users:
The GVE is used by U.S.A. government contractors on security-related projects.

Applications:

- M.K. Smith, A. Siebert, B. Divito, D. Good. *A Verified Encrypted Packet Interface.* Software Engineering Notes 6,3 (July 1981).

- D.I. Good, A.E. Siebert, L.M. Smith. *Message Flow Modulator - Final Report.* Technical Report 34, Institute for Computing Science, The University of Texas at Austin, December 1982.

- J. Keeton-Williams, S.R. Ames, B.A. Hartman, R.C. Tyler. *Verification of the ACCAT-Guard Downgrade Trusted Process.* Technical Report NTR-8463, The Mitre Corporation, Bedford, MA, 1982.

- *Formal Specifications for Secure Communications Processor Trusted Software Release 2.2.* Honeywell Information Systems, Inc. March, 1985 (This was the first system awarded an A1 certification by the USA NCSC.)

- *Honeywell Logical Co-Processor Kernel (LOCK)* O.S. Saydjari, J.M. Beckman, J.R. Leaman. LOCKing Computers Securely. In Proceedings of the 10th National Computer Security Conference. Baltimore, MD, 1987.

- J. Freeman, R. Neely, L. Megalo, M. Krenzin. *Multinet Gateway Certification Program.* Final Technical Report, Ford Aerospace Corporation, RADC-TR-89-6, April 1989.

Availability:
Distribution of the GVE requires approval from the USA National Computer Security Center. Direct requests for the GVE (and the prerequisite government approval) to

Laura Tice
Computational Logic, Inc.
1717 W. 6th St., Suite 290
Austin, Texas 78703
phone: (512) 322-9951
email: tice@cli.com
fax: (512) 322-0656

Additional remarks:
None.

HOL

Participants (project leader, group members):
Worldwide

Survey contact:
Europe:
Dr. Mike Gordon
University of Cambridge
Computer Laboratory
New Museums Site
Pembroke Street
Cambridge CB2 3QG, UK.

North America:
Professor Phil Windley
Department of Computer Science
University of Idaho
Moscow, ID 83843, USA.
email: windley@cs.uidaho.edu

Level of effort (person-years, duration):
Continuous development since mid 1970s

Description:
The HOL system is an interactive theorem-proving environment for higher order logic. It is a development of Cambridge LCF which, in turn, is directly based Milner's original LCF system developed at Edinburgh.

HOL provides tools to support the proof of theorems in higher order logic. These tools are functions in the programming language ML, which forms the interface to HOL. It is intended that users of HOL will build their own application-specific theorem-proving infrastructure. ML is designed for this purpose.

The original HOL system (called HOL88) is fully implemented and available now. It is in the public domain and can be obtained by anonymous FTP.

HOL88 is implemented in Lisp and runs on any platform that supports Common Lisp (e.g. Suns, MIPS, HP workstations, Apple Macintosh, IBM PC). Two new versions of HOL, both implemented in Standard ML, are available: HOL90 from the University of Calgary is a public domain system; ICL HOL is a commercial system intended to support applications in the security critical area.

HOL is completely documented. There is a detailed description of the system, which includes a formal semantics of the version of higher order logic used, a manual for the ML programming language and a description of the theorem proving infrastructure. There is also a reference manual that documents every ML function in HOL. The text of this manual can be accessed by the help system and an X-windows browsing tool. Finally, there is a tutorial introduction to the system and a training course (including exercises and solutions). All the documentation is public domain and the LaTeX sources are distributed with the system.

Accomplishments:

Published articles or reports:

- M.J.C. Gordon, *HOL: A Proof Generating System for Higher-Order Logic*, in "VLSI Specification, Verification and Synthesis", edited by G. Birtwistle and P.A. Subrahmanyam, Kluwer, 1988.

- M.J.C. Gordon, *Mechanizing Programming Logics in Higher Order Logic*, in "Current Trends in Hardware Verification and Automated Theorem Proving", edited by G. Birtwistle and P.A. Subrahmanyam, Springer-Verlag, 1989.

- Phillip J. Windley, Myla Archer, Karl N. Levitt, and Jeffrey J. Joyce (eds), *Proceedings of the Third International Workshop on the HOL Theorem Proving System and its Applications*, IEEE Computer Society Press, 1992.

- HVG Cambridge, *The HOL System: TUTORIAL, DESCRIPTION, REFERENCE* (three volumes). Produced by the Cambridge Research Centre of SRI International with the support of DSTO Australia. Distributed with the HOL88 System (Version 2.0 or later).

Status:
Public domain and free

Start date:

Completion:

Future developments:
The core system is complete and stable. Current work (supported by DSTO Australia and in collaboration with Inria, France) aims to build a new colour graphic interface and to develop a methodology for its use. This work has been in progress for about a year.

Libraries of theories and theorem proving tools are provided by the user community and are distributed with new releases of the system.

Strengths, weaknesses and suitability:

HOL is particularly suitable as a basis for building highly secure application-specific verification environments. It lacks some features that are built in to other theorem provers, such as fast decision procedures. It thus requires more manual guidance than these. However, it is expected that such productivity-enhancing tools will eventually be provided as contributed libraries.

External users:

A list of ongoing projects with HOL is maintained as part of the system documentation. The version of this list in the current release includes projects from the following organizations.

IMEC, University of Calgary, University of Cambridge, Syracuse University, University of British Columbia, Mitsubishi Research, ICL Secure Systems, University of Idaho, Digital, IBM, Queens University, Hewlett Packard, Boeing, Amdahl, University of California at Davis, SRI Cambridge Research Centre, Inmos Limited, British Telecom, University of Warwick, British Aerospace, University of Oxford, Abo Akademi (Finland), University of Cape Town, DSTO Australia, University of Washington, TFL Denmark.

N.B. There are users of HOL not in this list (and probably some of those in the list no longer use HOL).

There is an annual International HOL Users Meeting (there have been four so far). These are expected to be alternatively in Europe and North America. The Meeting in 1991 was at the University of California at Davis.

Applications:

Mathematical proof checking, hardware, software and system verification.

Availability:

On tape or by anonymous FTP from Phil Windley (windley@cs.uidaho.edu) or Cambridge University (hol-support@cl.cam.ac.uk).

Additional remarks:

ICL HOL

Participants (project leader, group members):

ICL HOL is system supporting formal specification and verification. It is being developed by International Computers Limited as part of the FST Project (IED Project 1563). Other partners in this project are Program Validation Limited and the Universities of Cambridge and Kent. The project is jointly funded by the partners and the Information Engineering Directorate of the UK Department of Trade and Industry.

Survey contact:

Roger Jones,
ICL Associated Services Division,
Eskdale Road,
Winnersh,
Wokingham,
Berkshire RG11 5TT, UK

Tel: +44 734 69 3131
email: R.B.Jones@win0109.uucp

or

Geoff Scullard,
ICL,
Wenlock Way,
Manchester M12 5DR,
UK
Tel: +44 61 223 13 01
email: G.T.Scullard@win0109.uucp

Level of effort (person-years, duration):
3-4 man years.

Description:
ICL HOL is a re-engineering of the well established Cambridge HOL system, and has as its objectives:

- product quality implementation and documentation;
- high assurance in the integrity of the proof system, i.e., a high degree of confidence in the validity of the theorems it is used to prove;
- improved ease of use, for the novice user, for the expert and for the implementer of extensions to the system;
- good facilities for extending the system to provide proof support for formalisms other than HOL, including other specification and programming languages.

The quality and integrity of the system are assured by the use of good engineering practice and, in particular, the LCF paradigm found in the Cambridge system. Thus a logical kernel of code, containing protected datatypes, supplies the only means of generating theorems. The primitive inference rules implemented in the logical kernel are based on a fully formal specification of the HOL language and logic (which is available on request from ICL). The protection mechansisms used in the kernel, e.g., the means whereby the user is prevented from using a theorem in an inappropriate context, have also been formally specified in detail. All features of the Cambridge system which, through accident or misuse, could allow fallacies to be proved, have been omitted.

Another facet of the LCF paradigm is that the user works within the context of a metalanguage. This provides the user with access to a typed functional programming language and most of the system builder's tools, while the type system of the language prevents theorems being created other than through the logical kernel. This makes the resulting systems almost as extensible by the user as by the system builder, which can be a great benefit to the experienced user. The drawback is that this approach gives the user access to a vast number of functions, though some are aimed only at system extenders. These include long lists of derived inference rules and tactics. One aim of the new system is to organise and present this material so as to make it usable, while not providing any functions that are redundant. Apart from the obvious provision of tutorial and reference material three other methods have been used:

- the use of naming conventions, covering suffixes, prefixes and case of letters;

- comprehensive coverage - for instance the inference rules provided include all the rules given in a standard presentation of the predicate calculus;

- consistent coverage - if a pattern of coverage emerges then it is followed to its conclusion.

Accomplishments:

The system has been used to prototype a proof support system for Z, including syntax checking, type inference and a selection of proof rules covering set theory and predicate calculus. A more complete implementation is planned for late 1992.

Other recent applications are an experimental implementation of the BAN authentication logic and a treatment of the possible worlds semantics for propositional modal logic.

Published articles or reports:

- R.D. Arthan. *Formal Specification of a Proof Tool.* In S.Prehn and W.J.Toetenel, editors, VDM '91, Formal Software Development Methods, LNCS 551, volume 551, pages 356–370. Springer-Verlag, 1991.

- R.D. Arthan. *A Report on ICL HOL.* In proc. 1991 HOL Workshop. ACM/IEE (to appear), 1992.

- K. Blackburn. *A Report on ICL HOL.* Lecture Notes in Computer Science, 1992.

- R.B. Jones. *Methods and Tools for the Verification of Critical Properties.* In R.Shaw, editor, proc. 5th BCS-FACS refinement workshop. Springer-Verlag, 1992.

- The following internal reports can be made available on request:

- DS/FMU/IED/SPC001. *HOL Formalised: Language and Overview.* R.D. Arthan, ICL Secure Systems, WIN01.

- DS/FMU/IED/WRK020. *Implementing the Authentication Logic.* D. J. King, ICL Secure Systems, WIN01.

- DS/FMU/IED/WRK022. *Modal Logic in HOL.* R.D. Arthan, ICL Secure Systems, WIN01.

Status:

Field trials during 1992.

Start date:

Research and prototyping - late 1989.
Product development - early 1991.

Completion:
   Ongoing.

Future developments:
   Research is continuing into proof automation, HCI, proof support for Z, and verification of critical aspects of the logical kernel.

Strengths, weaknesses and suitability:
   The system is particularly applicable in areas where the integrity of proof is of overriding importance or as a generator of reliable proof support environments for other languages.

External users:

Applications:
   The system is intended to support highly assured applications in secure and safety-critical systems.

Availability:
   To be published in the ICL HOL Newsletter (on request from the above).

Additional remarks:
   At present, the system runs on Sun 3 or Sun 4 and requires the Poly/ML compiler from Abstract Hardware Limited.

HP-SL

Participants (project leader, group members):
   HP Laboratories, Software Engineering Department

Survey contact:
   Patrick Goldsack
   Software Engineering Department
   HP Laboratories
   Filton Road
   Bristol BS12 6QZ

   email: pcg@hplb.hpl.hp.com.uk
   tel: +44-272-799910 x 28146
   fax: +44-272-228003

Level of effort (person-years, duration):
   3 person years, ongoing.

Description:
   HP-SL is a formal specification notation based loosely on VDM/Z, but with several key differences, e.g.

   - polymorphism

   - uses a higher-order logic

   - it has a module system with module combinators such as enrichment, summation and parameterisation

The tools provide the basic facilities of formatting and type-checking, working equally with partial specifications and specifications spread throughout specification documents and surrounded by text. These tools may be used either batch, or interactively through an editor interface. Extensive specification browsing and indexing facilities are also provided.

Accomplishments:

HP-SL has been used on a series of HP product developments, see the description of the Hewlett Packard Leadership Projects in application survey.

Tools have been demonstrated at the VDM '90 and VDM '91 conferences.

Published articles or reports:

- S. P. Bear, *An Overview of HP-SL*, in proceeding of VDM '91: Formal Software Development Methods, The Netherlands, October 1991, Lecture Notes in Computer Science 551, Springer-Verlag 1991.

- C. Dollin, *The HP-ST Toolset*, in proceeding of VDM '91: Formal Software Development Methods, The Netherlands, October 1991, Lecture Notes in Computer Science 551, Springer-Verlag 1991.

- P. C. Goldsack, *A tour of HP-SL*, Hewlett Packard Laboratories Technical Memo HPL-91-68, 1991.

- P. C. Goldsack, *HP-SL concrete syntax*, Hewlett Packard Laboratories Technical Memo HPL-91-74, 1991.

- P. C. Goldsack, *HP-SL abstract type syntax*, Hewlett Packard Laboratories Technical Memo HPL-91-69, 1991.

- P. C. Goldsack, *The HP-SL model of polymorphism*, Hewlett Packard Laboratories Technical Memo HPL-91-71, 1991.

- P. C. Goldsack, *The HP-SL type model*, Hewlett Packard Laboratories Technical Memo HPL-91-73, 1991.

- P. C. Goldsack, *HP-SL pre-defined types*, Hewlett Packard Laboratories Technical Memo HPL-91-72, 1991.

- P. C. Goldsack, *Module Combinators for Specification Languages*, Hewlett Packard Laboratories Technical Memo HPL-91-70, 1991.

Status:

Work is on-going - currently we have

- completed the initial tools - these are available on beta-release
- completed the training material
- extensive language documentation, though we still require the completion of the proof rules documentation

Start date:

1988

Completion:

No fixed completion date.

Future developments:
Completion of language description. Additional tool support for manipulating and exploring specifications.

Strengths, weaknesses and suitability:

Strengths The notation is small, very regular and extremely expressive. This results in a notation which is relatively easy to learn, to tool and to use.

The tools have simple interface that provides an interactive environment enabling users to carry out most of the required tasks in producing specification documents.

Weaknesses HP-SL is currently lacking some documentation - in particular a document describing the proof theory, and a tutorial document. Work is in progress on these.

HP-SL has no explicit concurrency primitives - though we have developed styles of use of HP-SL to cover areas which include concurrency and real-time performance issues.

There are no tools to support reasoning or transformation, etc., though some of these are under development. The tools do not provide for versioning, configuration, tracability etc.

Applications/Suitability We have sucessfully used the notation and tools internally since 1988 on real product developments. These products have been from such diverse areas as CAD systems, medical systems (including real-time) and software engineering environments.

External users:
The tools have been licenced to over 20 universities around the world. Some of these are currently evaluating HP-SL for use on formal specification courses.

Applications:

Availability:
A one-year beta licence is available for evaluation purposes. Longer licences can be arranged if there is a specific need. The software is available for HP and Sun workstations.

Additional remarks:
None

The IPTES VDM-SL Toolbox

Participants (project leader, group members):
The Institute of Applied Computer Science, Denmark

Survey contact:
Poul Lassen,
The Institute of Applied Computer Science,
Forskerparken 10,
DK-5230 Odense M,

Denmark
e-mail: iptes@ifad.dk

Level of effort (person-years, duration):
3 man years are used in the development of the Toolbox.

Description:
A number of tools which support the specification, execution and analysis of an executable subset of BSI VDM-SL. The IPTES VDM-SL toolset includes tools for syntax checking, interpretation, pretty printing, static semantic analysis and debugging of specifications written in the standard ASCII syntax of an executable subset of BSI VDM-SL. The executable subset supported by the toolbox is exceptional by including loose specifications, very strong pattern matching facilities and imperative as well as applicative parts of the BSI VDM-SL language.

Accomplishments:

Published articles or reports:

- Peter Gorm Larsen, Poul Boegh Lassen. *An Executable Subset of Meta-IV with Loose Specification.* In VDM'91 Symposium, VDM Europe, Springer-Verlag, March 1991.

- Michael Andersen, Rene Elmstroem, Poul Boegh Lassen, Peter Gorm Larsen. *Making Specifications Executable Using IPTES Meta-IV.* Accepted for Euromicro'92 Conference, September 1992.

Status:
Currently a syntax checker, interpreter and ASCII to LAT$_E$X generator is completed. The static semantic analysis and debugger is currently being implemented.

Start date:
January 1991

Completion:
Static Semantic Analyzer: July 1992, Debugger: July 1992.

Future developments:
Besides from the above mentioned extensions there are plans to include support for Looseness Analysis (calculating all legal results of a loose specification), Heterogeneous execution (allowing to execute specifications where part are implemented as C++ code) and code generation (generating C++ code from IPTES VDM-SL specifications).

Strengths, weaknesses and suitability:
The strength of the IPTES VDM-SL language compared to other executable subsets of VDM-SL is the support for loose specification, strong pattern matching and inclusion of imperative as well as applicative constructs. The strength of the interpreter is the short turn-around time for executing and changing specifications compared to most executable subsets of VDM-SL where execution is provided through compilation.

External users:
In ESPRIT project EP5570 several tools of the toolbox are used. Users in

this project are: Telefonica I+D, Spain, Commissariat a l'Energie Atomique, France and ENEA, Italy.

Applications:
Debugger, semantic analyzer.

Availability:
Pre-release of the toolbox including, parser/scanner, pretty printer and interpreter for the IPTES VDM-SL language is available for SUN4 workstations for a handling fee of 550 Dollars US.

Additional remarks:

IBM Z tool

Participants (project leader, group members):
IBM United Kingdom Laboratories Ltd

Survey contact:
Jonathan Hoare,
IBM United Kingdom Laboratories Ltd,
Hursley Park,
Winchester, SO21 2JN

Level of effort (person-years, duration):
5 man-years to develop, half person to maintain/enhance

Description:
Live parsing editor for Z documents based on LPEX, an editor for OS/2.
Facilities for accelerated editing.
Syntax, scope, and type checking.
Schema expansion.
Cross referencing of Z documents.

Accomplishments:
Successfully used by designers of CICS/ESA at Hursley with considerable productivity improvements.

Published articles or reports:
None to date.

Status:
Complete - in maintenance.

Start date:
Jan 1990

Completion:
Jan 1991

Future developments:
Maintenance/enhancement

Strengths, weaknesses and suitability:

Strengths Flexible friendly editing environment.

Weaknesses Specific to IBM's BookMaster Generalised Markup Language, Needs PS/2, OS/2 and LPEX.

External users:
None.

Applications:
Improves productivity in Z specification and design.

Availability:
IBM internal use only

Additional remarks:
Demonstrated at external conferences and seminars: The Z User Meeting, VDM '91, etc.

Jape

Participants (project leader, group members):
Oxford University Computing Laboratory and
Queen Mary and Westfield College, London University

Survey contact:
Bernard Sufrin (sufrin@uk.ac.ox.comlab)
Richard Bornat (richard@uk.ac.qmw.csd)

Level of effort (person-years, duration):

Description:
Jape is an interactive tool designed to help both the teaching and the application of formal reasoning.

Jape does not have a preference for any particular logic, but just embodies a theory of hypothetical deduction, and a theory of safe substitution. It has been applied to several logics, including predicate calculus, operational semantics, Milner-style type theory, typed set theory (Z), a functional programming logic, and a Hoare style logic of program refinement.

Any formal system whose view of safe substitution is consistent with Jape's can be described simply in the Jape metalanguage. Inference systems (such as those used to present the operational semantics of languages with scope) in which the "hypothesis" is rather more structured than just a set of logical formulae are also accomodated.

Support for reasoning in the equational style in an "open" tool such as Jape (which knows nothing, a priori, about equality) presents some interesting problems which we believe that we have solved by providing the means to elide material from the presentation of proof trees. The same methods are used to support program refinement proofs.

User interaction with the tool during a proof is easy, and for certain logics (predicate calculus included) it can be spectacularly easy! During proofs in these logics the proof tree is directly manipulated during the proof simply by clicking on a hypothesis or conclusion with the mouse.

The logic designer uses a language of tactics (which may be bound to mouse gestures, menu items, keystrokes) to achieve the desired interface between user and tool. The most basic tactics are the rules of inference of the object logic, and there is small, carefully selected, collection of methods of composing tactics into strategies which support the discovery of proofs.

Accomplishments:

We have begun to use the tool (on a very small scale) to help students gain experience with natural deduction proofs in predicate calculus, to introduce polymorphic type inference systems, and to explore some simple ideas in functional programming.

Published articles or reports:

None, as yet.

Status:

Start date:

March 1992

Completion:

The present prototype has been under development since March 1992 and has been through several stages of development. The project is open-ended.

Future developments:

Strengths, weaknesses and suitability:

Strengths Interaction during a proof is easy – the tool doesn't impose any artificial barriers to proof (if the problem is hard, or the logic is arcane then the tool won't make proof any easier, of course!)

The tool tracks (and polices, insofar as it is possible) the syntactic side-conditions of rules.

Being based on unification (rather than pattern matching) means that the tool may be used in calculation as well as in proof.

The learning curves for both users and logic designers are rather shallow. One doesn't have to learn a programming language in order to be able to do proofs. Moreover using the metalanguage to describe a logic is rather easy. For example the following rules are extracted from a presentation of predicate logic:

RULE $\wedge \vdash$ (P,Q,R) FROM P, Q $\vdash$ R INFER P $\wedge$ Q $\wedge$ R

RULE $\vdash \wedge$ (P,Q,R) FROM $\vdash$ P AND $\vdash$ Q INFER $\vdash$ P $\wedge$ Q

RULE specialise(E,xs,P,C) WHERE VAR xs AND SAFE P[xs \ E] FROM $\vdash$ forall xs . P INFER $\vdash$ P[xs \ E]

Weaknesses The tool is based on an extensible language of formulae so if questions of syntactic "well-formedness" of object language formulae are important, they must be axiomatised as part of the logic rather than simply defined as part of the language's grammar. In practise this weakness hasn't proven to be too onerous.

In the present prototype there is little or no support for proof activity which spans more than one session with the tool.

Suitability We believe that at present the tool may usefully be used to help train people in proof methods.

External users:

Applications:
(see description)

Availability:
(alpha test versions may be available to selected collaborators from October 1992)

Additional remarks:

LOEWE

Participants (project leader, group members):
Günter Karjoth, Carl Binding, Jan Gustafsson.

Survey contact:
Günter Karjoth,
IBM Research Division,
Zurich Research Laboratory,
Säumerstr. 4,

8803 Rüschlikon,
Switzerland
Tel: +41 (1) 7248-486
Email: gka@zurich.ibm.com

Level of effort (person-years, duration):
Approximately 7 person-years

Description:
LOEWE is a prototype of an integrated tools environment for the specification, analysis, and implementation of communication software. Although primarily based on the formal description technique LOTOS, LOEWE additionally offers multiple, semantically equivalent representations of a given protocol specification. Each representation is specially adapted to support a specific class of tools: *processes* are used for simulation and compilation, fast state space exploration is performed on *extended finite state machines*, and *labelled transition graphs* are used for system verification.

LOEWE currently contains a LOTOS syntax and static semantic verifier, a generator of extended finite state machines with associated interactive state exploration tools, a compiler to generate C code from LOTOS behavior expressions, and a translator of LOTOS abstract data type operations to a set of LISP functions.

LOEWE incorporates tools from other sites; specifically these are a temporal logic model checker and a branching bisimulation tool.

Accomplishments:
LOEWE's translation algorithms generate a labelled transition graph of a finite state LOTOS specification thus enabling the application of a number of effective analyzing methods. Several state graphs have been generated from

public LOTOS specifications. The largest graph comprised about 1 million states, surpassing any result reported in literature. Thus far, LOEWE also produced smaller state graphs for the same LOTOS specification.

Published articles or reports:

- G. Karjoth, C. Binding, and J. Gustafsson. *LOEWE: A LOTOS engineering workbench.* Research Report RZ 2143 (#74743), IBM Research Division, 06/17/91. A revised version will appear in *Computer Networks and ISDN Systems*: Special issue on "Tools for FDTs", North-Holland.
- G. Karjoth. *Implementing LOTOS specifications by communicating state machines.* Research Report RZ 2283 (#76813), IBM Research Division, 03/02/92.

Status:
LOEWE prototype exists.

Start date:
1988

Completion:
Ongoing research, development and application.

Future developments:
Continued improvement of the system's performance. Incorporation of methods combating the state explosion problem, e.g. symbolic representations of the state space by binary decision diagrams.

Strengths, weaknesses and suitability:
The transformation of the original LOTOS specification into derived representations allows the construction of efficient validation tools (state exploration, model checking) and can also be used to generate a directly executable implementation of the original specification. For the latter, the use of an intermediate execution model in the form of a network of extended finite state machines supports the generation of highly optimized code.

Several of the representations used are not LOTOS-dependent, but are more general in nature. They can be used in workbenches supporting other formalisms or simply in special-purpose tools for a given functionality, such as for state exploration. Not only CCS-type languages can be used as "front-ends" to such systems, but also languages based on asynchronous communication such as Estelle and SDL could possibly profit from our extended finite state machine model.

Although LOTOS specifications of medium size have been handled with success, further improvements and extensions are necessary in order be able to treat specifications originating from the standards bodies.

External users:
None at this stage.

Applications:

Availability:
It has not yet been released.

Additional remarks:
Supported in part by the RACE project N° 1046, Specification and Programming Environment for Communication Software (SPECS).

LOLA (LOtos LAboratory)

Participants (project leader, group members):
Juan Quemada, Santiago Pavon, David Larrabeiti, Martin Llamas, Maria Hultstrom, Angel Fernandez, Jose A. Manas.

Survey contact:
Santiago Pavon
email: spavon@dit.upm.es

Juan Quemada
email: jquemada@dit.upm.es

Dpto. Telematica, E.T.S.I. Telecomunicacion,
Ciudad Universitaria s/n, E-28040 MADRID, SPAIN
Tel: +34 1 5495700 ext: 375, 368
telefax: +34 1 5432077
telex: 47430 ETSIT E

Level of effort (person-years, duration):
Six man-years

Description:
LOLA is a LOTOS state exploration tool which uses state space compression techniques in order to improve its efficiency. It has been conceived using a transformational approach.

The transformations of LOLA automate different design tasks: simulation and transformation of the LOTOS specification, calculation of the transition system generated by a LOTOS specification, calculation of the response of a specification to a test (it follows the definition of the Testing Equivalence), symbolic execution of the LOTOS specifications, deadlock detection, deriving efficient implementations, etc.

LOLA supports full LOTOS language. However, the data values are treated operationally by interpreting equations as rewrite rules from left to right. Therefore LOLA does not process LOTOS, but rather an operational version of it where equations, interpreted from right to left, are a set of rewrite rules.

Accomplishments:

Published articles or reports:

- *The Testing Functionalities of LOLA* S. Pavon, M. Llamas, Formal Description Techniques III, FORTE-90, pp. 559-562, North-Holland, 1991.

- *Contribucion al Analisis y Transformacion de Especificaciones LOTOS* Santiago Pavon, E.T.S.I. Telecomunicaion, Universidad Politecnica de Madrid, Phd. Thesis 1991.

- *State Exploration by transformation with LOLA* J. Quemada, S. Pavon, A. Fernandez, Automatic Verification Methods For Finite State Systems, pp. 294-302, Springer–Verlag, 1990.

- *Transforming LOTOS Specifications With Lola: The Parameterized Expansion* J. Quemada, S. Pavon, Formal Description Techniques, FORTE-88, pp. 45-54, North-Holland, 1989.

- *LOLA: Design and Verification of Protocols using LOTOS* J. Quemada, A. Fernandez, J.A. Manas, In proceedings: Ibericom, Conf. on Data Communications, Lisbon, 1987.

Status:
Version 2R1 is ready for distribution.

Start date:
1985

Completion:
Last version: 1992. Ongoing as a research project.

Future developments:
New state exploration functions with applications in testing and in efficient implementation derivation. Protocol synthesis and functionality decomposition.

Strengths, weaknesses and suitability:

External users:

Applications:

Availability:
ftp 138.4.2.2 (login:anonymous, chdir to info/src/lotos) or contact with lotos@dit.upm.es

Additional remarks:

LP

Participants (project leader, group members):
Steve Garland, John Guttag (MIT), Jim Saxe, Jim Horning (DEC Systems Research Center)

Survey contact:
John Guttag
MIT Laboratory for Computer Science
545 Technology Square
Cambridge, MA 02139
U.S.A.

Level of effort (person-years, duration):
6-8 person-years to date

Description:

Proof debugger for a subset of first-order logic, intended for use in conjunction with other Larch tools.

Accomplishments:
Verifications of non-trivial software and hardware designs (see publications).

Published articles or reports:

- Stephen J. Garland and John V. Guttag *A Guide to LP, The Larch Prover* DEC/SRC Report 82, 1991

- S.J. Garland, J.V. Guttag, and J.J. Horning *Debugging Larch Shared Language Specifications* IEEE TSE, Vol. 16, No. 9, 1990 [This is a revised and expanded version of S.J. Garland and J.V. Guttag *Using LP to Debug Specifications* Elsevier, Proc. IFIP TC-2 WC on Programming Concepts and Methods, 1990]

- J.B. Saxe, S.J. Garland, J.V. Guttag, and J.J. Horning *Using Transformations and Verification in Circuit Design* DEC/SRC Report 78, 1991 also Proceedings of the Workshop on Designing Correct Circuits, North-Holland, 1992

- S.J. Garland, J.V. Guttag, and J. Staunstrup *Verification of VLSI Circuits using LP* Proceedings of IFIP WG 10.2, The Fusion of Hardware Design and Verification, North Holland, 1988

- J. Staunstrup, S.J. Garland, and J.V. Guttag *Localized Verification of Circuit Descriptions* Proc. Workshop on Automatic Verification Methods for Finite State Systems Lecture Notes in Computer Science, Springer-Verlag, 1989

- U. Martin and T. Nipkow *Automating Squiggol* Proc. IFIP TC-2 WC on Programming Concepts and Methods, 1990

- J.M. Wing and C. Gong *Experience with the Larch Prover* Proc. ACM Workshop on Formal Methods in Software Development, 1990

- C. Gong and J.M. Wing *Raw Code, Specification, and Proof of the Avalon Queue Example* CMU Report CMU-CS-89-172, 1989

- J.M. Wing and C. Gong *Machine-Assisted Proofs of Properties of Avalon Programs* CMU Report CMU-CS-89-171, 1989

Status:
Release 2.2 available now; development continuing

Start date:
1987 (evolved from earlier tool, REVE)

Completion:
Ongoing

Future developments:
Full quantifiers

Strengths, weaknesses and suitability:
Uses classical proof techniques, term rewriting for simplification, case analysis, induction, contradiction, etc.

Usually fails quickly (and often informatively) on non-proofs. Designed for interactive proof development and non-interactive replay during proof maintenance (regression testing)

External users:
Not stated.

Applications:
Software and hardware verification

Availability:
Anonymous ftp from larch.lcs.mit.edu; free

Additional remarks:
Most mature of the Larch tools. Being used as proof engine for various front-end tools Synchronized Transitions (Staunstrup) TLA+ (Lamport) LSL (Horning)

The Boyer-Moore Theorem Prover (Nqthm)

Participants (project leader, group members):
Robert S. Boyer and J. Strother Moore

Survey contact:
J. Strother Moore
Computational Logic Inc.
1717 W. 6th St., Suite 290
Austin, TX 78703

phone: (512) 322-9951
email: moore@cli.com
fax: (512) 322-0656

Level of effort (person-years, duration):
2 persons for 21 years

Description:
This is an automatic/interactive theorem prover for a first order logic resembling Pure Lisp. The logic is based on recursive function definition and inductively defined data types. The theorem prover consists of about one megabyte of Common Lisp and contains many heuristics for controlling rule-based rewriting and induction.

Accomplishments:
Proofs of Goedel's Incompleteness theorem, Wilson's Theorem, Gauss' law of quadratic reciprocity, unsolvability of the halting problem, correctness of the Boyer-Moore fast string searching algorithm, invertibility of the RSA encryption algorithm, Piton assembler, Micro-Gypsy compiler, FM8502 microprocessor, FM9001 microprocessor, KIT operating system, Bitonic sort, several programs in Unity, asynchronous communications, biphase mark protocol, 8-bit parallel io Byzantine agreement processor, a formalization of the Motorola MC68020, a variety of MC68020 machine code programs, a mutual exclusion algorithm, and many other problems.

Published articles or reports:

- A Computational Logic, Boyer & Moore, Academic Press, 1979.
- A Computational Logic Handbook, Boyer & Moore, Academic Press, 1988.

Status:

Start date:
1971

Completion:
1992

Future developments:
A new theorem prover based on applicative Common Lisp with significantly greater control over hints, theories and libraries, an expanded set of built-in data types, random-access arrays, and many other improvements. A prototype currently exists.

Strengths, weaknesses and suitability:
Sound, extensible, interactive via incorporation of user-suggested but mechanically proved rules. Complicated; requires experience. Seems particularly suited for proofs about computer systems, but has been used successfully on a wide variety of applications.

External users:
Hundreds of copies have been distributed over the net. Mainly university research groups. Some industry interest. Much government interest.

Applications:
See "Accomplishments" above.

Availability:
To get a copy follow these instructions:

1. ftp to Internet host cli.com.
   (cli.com currently has Internet number 192.31.85.1)

2. log in as ftp, password guest

3. get the file /pub/nqthm/README

4. read the file README and follow the directions it gives.

Inquiries concerning tapes may be sent to: Computational Logic, Inc., Suite 290, 1717 W. 6th St., Austin, Texas 78703.

Additional remarks:
See also the corresponding response for Pc-Nqthm.

This development of Nqthm has been an ongoing effort for 21 years. It has received support from a wide variety of sponsors including Science Research Council of Great Britain (now Science and Engineering Research Council), Xerox, SRI International, National Science Foundation, Office of Naval Research, NASA, Air Force Office of Scientific Research, Digital Equipment Corporation, the University of Texas at Austin, the Venture Research Unit of

British Petroleum, Ltd., IBM, Defense Advanced Projects Research Agency, National Computer Security Center, the Space and Naval Warfare Systems Command, and Computational Logic, Inc. Our primary support now comes from Defense Advanced Research Projects Agency, DARPA Order 7406. The views and conclusions contained in this document are those of the author(s) and should not be interpreted as representing the official policies, either expressed or implied, of Computational Logic, Inc., the Defense Advanced Research Projects Agency or the U.S. Government.

Pc-Nqthm (An interactive "Proof-checker" enhancement of the Boyer-Moore Theorem Prover

Participants (project leader, group members):
Matt Kaufmann

Survey contact:
Matt Kaufmann
Computational Logic Inc.
1717 W. Sixth St., Suite 290
Austin, TX 78703

phone: (512) 322-9951
email: kaufmann@cli.com
fax: (512) 322-0656

Level of effort (person-years, duration):
Perhaps (very roughly) 2 man-years, spread over 5 years.

Description:
This "proof-checker" is loaded on top of the Boyer-Moore Theorem Prover; see the corresponding response for Nqthm. The user can give commands at a low level (such as deleting a hypothesis, diving to a subterm of the current term, expanding a function call, or applying a rewrite rule) or at a high level (such as invoking the Boyer-Moore Theorem Prover). Commands also exist for displaying useful information (such as printing the current hypotheses and conclusion, displaying the currently applicable rewrite rules, or showing the current abbreviations) and for controlling the progress of the proof (such as undoing a specified number of commands, changing goals, or disabling certain rewrite rules). A notion of "macro commands" lets the user create compound commands, roughly in the spirit of the tactics and tacticals of LCF and its descendents. An on-line help facility is provided, and a user's manual exists.

As with a variety of proof-checking systems, this system is goal-directed: a proof is completed when the main goal and all subgoals have been proved. Upon completion of an interactive proof, the lemma with its proof may be stored as a Boyer-Moore "event" that can be added to the user's current library of definitions and lemmas. This event can later be replayed in "batch mode". Partial proofs can also be stored.

Accomplishments:
This system has been used to check theorems stating the correctness of a transitive closure program, a Towers of Hanoi program, a ground resolu-

tion prover, a compiler, irrationality of the square root of 2, an algorithm of Gries for finding the largest "true square" submatrix of a boolean matrix, the exponent two version of Ramsey's Theorem, the Shroeder-Bernstein theorem, Koenig's tree lemma, and others. It has also been used to check the correctness of several Unity programs and has been used for hardware verification.

Published articles or reports:
The first one below is a detailed user's manual, including soundness arguments. The second extends this by describing an extension of the system which admits free variables, an important addition for doing full first-order reasoning. The third is a reference for that full first-order reasoning capability.

- Matt Kaufmann, *A User's Manual for an Interactive Enhancement to the Boyer-Moore Theorem Prover*. Technical Report 19, Computational Logic, Inc., May, 1988.

- Matt Kaufmann, *Addition of Free Variables to an Interactive Enhancement of the Boyer-Moore Theorem Prover*. Technical Report 42, Computational Logic, Inc., May, 1989.

- Matt Kaufmann, *An Extension of the Boyer-Moore Theorem Prover to Support First-Order Quantification*. To appear in J. of "Automated Reasoning" A preliminary (and expanded) version appears as Technical Report 43, Computational Logic, Inc., May, 1989.

Status:
Ongoing

Start date:
Late 1986

Completion:
1992

Future developments:
Boyer and Moore are creating a new theorem prover based on applicative Common Lisp; see the "Future developments" response for Nqthm. There currently exists a similar interactive enhancement for that system, and we intend to continue development of that capability.

Strengths, weaknesses and suitability:
Strengths

- Combination of capability for high degree of user control with the power of the Boyer-Moore prover
- On-line help facility and users manuals
- Extensibility by way of "macro commands" (patterned after the tactics and tacticals of LCF, HOL, Nuprl etc.)
- Full integration into Boyer-Moore system
- Careful attention to soundness issues

Weaknesses

- Ease of low-level interaction often tempts users to construct ugly proofs without many reusable lemmas that are hard to modify.

- First-order quantification is handled via Skolemization, rather than directly (as in the Never prover).

External users:
Dozens of copies have been distributed over the net. Mainly university research groups. Some industry interest. Much government interest.

Applications:
See "Accomplishments" above

Availability:
To get a copy, first obtain the Boyer-Moore Theorem Prover (Nqthm), as described in the corresponding response for Nqthm. Then follow these instructions:

1. ftp to Internet host cli.com. (cli.com currently has Internet number 192.31.85.1)

2. log in as ftp, password guest

3. get the file /pub/proof-checker/README-pc

4. read the file README-pc and follow the directions it gives.

Inquiries concerning tapes may be sent to: Computational Logic, Inc., Suite 290, 1717 W. 6th St., Austin, Texas 78703.

Additional remarks:
See also the corresponding response for Nqthm.

This work was supported by the Defense Advanced Research Projects Agency (currently DARPA Order 7406), the Office of Naval Research, the National Computer Security Center, and IBM. The views and conclusions contained in this document are those of the author and should not be interpreted as representing the official policies, either expressed or implied, of Computational Logic, Inc., the Defense Advanced Research Projects Agency or the U.S. Government.

Nuprl Proof Development System

Participants (project leader, group members):
Professors Robert L. Constable and Douglas J. Howe, Stuart Allen, William Aitken, James Caldwell, Rich Eaton, Paul Jackson, and Judith Underwood.

Survey contact:
Liz Maxwell,
Dept. of Computer Science,
Cornell University,
Ithaca, NY 14853.

E-mail maxwell@cs.cornell.edu.

Level of effort (person-years, duration):
14 people x 9 years, 4 undergraduate part-time x 3 years.

Description:
Nuprl is a system developed at Cornell which is intended to provide an environment for the solution of formal problems, especially thoses where computational aspects are important. Nuprl involves several innovations, the two most important of which are its logic, its proof editor, and metalanguage interface .

Nuprl 3.2 is the latest release of the Nuprl system that we distribute to other sites. To date, Nuprl has been distributed to roughly 65 sites. These sites obtained the system via direct request (not via public ftp) and paid a small distribution fee of $150.

Nuprl 4.0 is the newest version of the system that incorporates many new ideas on the editing, structure and display of mathematical expresions. It also incoporates a reflection mechanism of the kind we studied in our previous theoretical work. Nuprl 4.0 is currently being used only at Cornell, and will likely be released next year. The third system is a prototype implemented in SML (instead of Common Lisp) that is functionally comparable to Nuprl 4.0 but can easily be reconfigured to accommodate other logics.

Nuprl is described in further detail in the book *Implementing Mathematics with the Nuprl Proof Development System*, Prentice Hall, 1986.

Accomplishments:
Nuprl has been extensively used at Cornell over the last several years. Nine Ph.D. theses about Nuprl have been completed since 1986. Some of the research based on the system is described below.

- Howe used Nuprl in the solution of an open problem in the theory of programming languages. To date no one has discovered a conventional solution to this problem; it appears that machine assistance is essential.

- David Basin built environments within Nuprl for reasoning about recursive function theory, set graph theory and computer hardware. He also built a number of general purpose tools that aid the building theories. These include a rewrite package and routines involving high-order matching, building on work of Paulson and the experience of others.

- Howe and a student have been investigating the implementation of constructive real analysis. Their work resulted in the first formal proofs of significant theorems in higher constructive mathematics.

- In collaboration with Profs. Brown and Leeser of Cornell's Electrical Engineering department, Basin and DelVecchio used Nuprl as part of their research into hardware verification. The first result is the verification of a real circuit (the MAEC-Mantissa Adjuster and Exponent Calculator). DelVecchio completely specified and verified a component (about 5,000 transistors for Mantissa adjustment) of an integrated circuit being designed at Cornell for NASA to analyze date in real time from an optical space telescope.

- Howe used Nuprl to investigate the use of reflection in automating reasoning. This effort supplied evidence of the practicality of the approach.

- Constable and C. Murthy used the system to apply results from proof theory to automatically extract computational information from certain kinds of non-constructive arguments in classical mathematics. The method they used involved automating several proof transformation procedures. They applied these transformations to a Nuprl-formalization of a well-studied classical proof of a famous combinatorial theorem, and were the first to obtain the construction implicit in the classical argument.

Published articles or reports:

Status:

    Funded.

Start date:

Completion:

Future developments:

We are in the process of building a state-of-the-art system to manage software development. We plan to do this by integrating several components that we understand well from long experience. These include the programming language Standard ML (SML), a specification, a tactic-oriented theorem proving environment written in SML, a library management system based on the type theory and written in SML, and a specialized collection of tactics which support a forward chaining theorem prover interacting with a library. The resulting *software management system* will provide substantial automated support for the design, implementation, documentation, verification and maintenance of software systems. It will enable us to directly transfer to a practical computing environment the experience we have gained over the past decade in building and using verification systems. It will also enables it to explore new issues of information management and knowledge-based theorem proving resulting from our improved understanding of the role of the library in the software development process.

Strengths, weaknesses and suitability:

External users:

To date Nuprl has been distributed to over 65 sites outside of Cornell. With a few exceptions, these sites obtained the system via direct request (not via public ftp), and paid a $150 distribution fee.

We do not know how the system is being used at all of the sites; below are a few applications we do know about.

Applications:

- Proof planning, Edinburgh University.

- Program synthesis in Nuprl, University of Darmstadt.

- Theorem proving in category theory, CMU.

- The implementation of constructive algebra, University of Surrey.

- Developing a model of software evolution; classical and constructive proofs, University of Virginia.

- Hardware verification, Johns Hopkins.

- Honors course on type theory, University of New South Wales.

- Applying type theory to the construction of distributed systems, i.e. extract parallel programs from a proof of their correctness, University of Dortmund.

- Developing a functional language for the formal description of digital and analog electronic systems. Transformational design and proving the behavioral equivalence of descriptions, Bell Telephone Mgf. Co, Belgium.

Availability:
Contact: Liz Maxwell
Dept. of Computer Science
Cornell University
Ithaca, NY 14853
maxwell@cs.cornell.edu

Additional remarks:

Penelope (Ada Verification)

Participants (project leader, group members):
Wolfgang Polak, David Guaspari, Carla Marceau, C. Douglas Harper, Geoffrey Hird, Carl Eichenlaub, K.T. Narayana and Maureen Stillman (Project Manager)

Survey contact:
Maureen Stillman
ORA Corporation
301A Dates Drive
Ithaca, NY USA 14850-1313

Phone: (607) 277-2020
FAX: (607) 277-3206
Internet: maureen@oracorp.com

Level of effort (person-years, duration):
36 person-years over 6 calendar years

Description:
The goal of the project was to develop a sound mathematical basis for verifying sequential Ada programs and to implement it in an environment. The

user of the environment is able to develop the specification, code and proof of correctness concurrently. We have adopted the Larch two-tiered approach to specification. The approach we have taken to verification is to compute an approximation to the weakest liberal precondition associated with each control point of an annotated Ada program. These preconditions are defined by predicate transformers derived from a denotational description of Ada semantics. A proof editor and extensive simplification capabilities can be invoked to simplify preconditions as the program is developed, and to prove the remaining verification conditions.

Accomplishments:

We have developed a two-tiered specification language called Larch/Ada and an extensible technique for defining its semantics. Predicate transformers have been defined for our subset of (sequential) Ada. To demonstrate the validity of our techniques we have proved the soundness of these transformers for a small, but difficult subset. We have defined what it means for a program to satisfy its specification when a two-tiered approach to specification is used. The Penelope environment implements the basic control constructs (if statement, loops, exit statement), arrays and records, packages, global variables, user-defined exceptions, built-in arithmetic functions, and user-defined subprograms (including recursive programs). We have implemented a library facility which supports separate verification. Some constructs supported by the theory (e.g., go to and case statements, constraint checking) have not been implemented.

Published articles or reports:

- David Guaspari, Carla Marceau and Wolfgang Polak, *Formal Verification of Ada programs.* IEEE Proceedings on Software Engineering, September 1990, pp. 1058-1075.

- Carla Marceau and C. Douglas Harper, *An Interactive Approach to Ada Verification.* NCSC Proceedings, October 1989.

- Wolfgang Polak, *A Technique for Defining Predicate Transformers.* Available as ORA Corporation Technical Report 89-53, 1989.

Status:

We are working on the implementation of private types and generics. They will be delivered at the completion of the project. We are also formulating a formal basis for Ada tasking. Integration into a software engineering environment is currently being researched. PCTE, a software engineering environment standard backplane is the target. We are associated with the DARPA STARS project, whose goal is to provide a software engineering environment for all phases of software development in Ada. Formal methods and tools will be an important part of that environment.

The final deliverables will be delivered on September 19, 1992. It will include the implementation of libraries, generics and private types.

Start date:

September 1986.

Completion:

September 1992.

Future developments:
Future work will include the implementation of tasking. The theory for constraint checking has been completed, and we will work on the implementation.

Strengths, weaknesses and suitability:

Strengths Penelope allows the user to develop programs and proofs concurrently and to reuse proof steps as the program changes. For example, it is often trivial to reprove a subprogram when it is changed from one that may be invoked only on "good" input data, to one that signals "bad" data by raising an exception. Penelope can be used to prove Ada packages which are stored in a library. Support for separate verification allows reusable components to be verified.

Weaknesses Penelope is a research prototype, not a production quality system. Penelope does not support verification of full Ada; tasking is not supported.

External users:
Mainly U. S. Government agencies and defense contractors, including NSA, NASA and U. S. Air Force Rome Laboratory.

Applications:
We have used the Penelope system to formally specify and verify Bell-LaPadula security invariants for a small set of kernel subprograms from the interprocess communication code of the Army Secure Operating System (ASOS). We have also verified a reusable calendar package from a NASA library and a software implementation of an RS232 repeater.

Availability:
For information on availability or to be included on the mailing list, please contact Maureen Stillman at the above address.

Additional remarks:

Romulus

Participants (project leader, group members):
Ian Sutherland, Li Gong, Adam Weitzman, Allan Heff

Survey contact:
Ian Sutherland
ORA Corporation
675 Massachusetts Ave.
Cambridge, MA USA 02139-3009

Phone: (617) 354-8230
FAX: (617) 354-6593
Internet: ian@cambridge.oracorp.com

Level of effort (person-years, duration):
3 year project

Description:

Romulus is a system for proving critical properties about system designs. Systems are specified as processes in a process algebra. The process algebra is described within the Higher Order Logic (HOL) theorem prover. The critical properties are described as properties of processes within HOL. Romulus was formerly called Ulysses.

One of the principal features that Romulus has that other verification systems typically lack is domain specific support for writing the specification. Most existing verification systems supply the user with very general constructs for writing specifications (e.g., arbitrary pre- and postconditions for a program, or arbitrary invariants for an abstract state machine). It is up to the user of the system to make sure that he uses these constructs to correctly state what the system is supposed to do. Usually, only sophisticated users can be trusted to do this correctly given the generality of the specification constructs. In the case of properties like computer security and fault tolerance, the user must be sophisticated not merely in formal methods, but also in the relevant field (computer security, fault tolerance, or whatever).

Romulus aims to remedy this deficiency by providing a collection of *generic* formal properties that capture various significant aspects of specific domains, such as security, integrity, fault tolerance, and real-time response. While these generic properties cannot be taken to capture *all* the requirements on a system, they do capture significant aspects of the most critical requirements on systems. Romulus users can state a completely general specification for their systems, but this specification can make use of the generic properties which are built into Romulus.

Accomplishments:

Romulus currently has support for proving computer security, that is, proving that sensitive information cannot be inferred by entities that are not authorized to have it. The definition of computer security used by Romulus was derived by applying a nonprobabilistic version of information theory to processes. Romulus includes a number of automatic analysis procedures which reduce proofs of general information-theoretic security to simple statements about data types.

Published articles or reports:

- Daryl McCullough, *Specifications for Multi-Level Security and a Hook-Up Property.* In Proceedings of the IEEE Symposium on Security and Privacy, pp. 161-166, 1987.

- Tatiana Korelsky et al., *Ulysses: A Computer-Security Modeling Environment.* In Proceedings of the 11th National Computer Security Conference, September, 1988.

- Garrel Pottinger and James Hook, *Ulysses Theories: The Modular Implementation of Mathematics.* ORA Corporation Technical Report 89-5, February, 1989.

- Daryl McCullough, *A Hook-Up Theorem for Multi-Level Security.*

IEEE Transactions on Software Engineering, June, 1990.

Status:
Romulus is currently being extended to specify and prove other critical properties, including correct authentication, data integrity, fault tolerance, and timeliness.

Start date:
Current enhancement effort began September 1990.

Completion:
Current enhancement effort ends September 1993.

Future developments:
Incorporation of support for automatic analysis of designs for fault tolerance and real-time response. Incorporation of CASE technology to enhance usability. Incorporation of a capability to specify designs in the LOTOS language. Incorporation of a capability to do refinement in HOL. Incorporation of cryptographic logics for analyzing authentication protocols.

Strengths, weaknesses and suitability:

Strengths Analysis of security is fully formally based on information theory, so its analysis is more comprehensive. Automatic analysis procedures do much of the work of security analysis automatically. Supports specification in a process algebra (PSL) whose syntax resembles imperative programming languages, making it easier for "naive" users.

Weaknesses No facility for refining a high level specification to a lower level specification (this is being addressed currently). No facility for linking to an implementation language.

External users:
U. S. Air Force Rome Laboratory, U. S. Naval Research Laboratory, NSA, Secure Computing Technology Corp.

Applications:
Romulus has been applied to specify several significant designs for secure systems, including a secure guard which is being fielded at Military Airlift command and a fault tolerant secure file system.

Availability:
US government agencies and their contractors can obtain Romulus.

Additional remarks:

SPADE: An Environment for Software Process Analysis Design and Enactment

Participants (project leader, group members):
The SPADE project is being carried out at the Politecnico di Milano and CEFRIEL, guided by Prof. Carlo Ghezzi and Prof. Alfonso Fuggetta with the collaboration of Ph.D. students and master students.

Survey contact:
  Sergio Bandinelli
  CEFRIEL,
  Via Emanueli, 15,
  20126 Milano (Italy),
  Tel.: +39-2-66100083,
  Fax: +39-2-66100448,
  E-Mail: bandinelli@mailer.cefriel.it.

Level of effort (person-years, duration):

Description:
  SPADE [BFG92,FG92] aims at the analysis design and enactment of software processes. It includes a graphical and textual language, called SLANG (Spade LANGuage) that provides suitable abstractions for the description of the process by which the software is produced. The graphical notation is based on ER nets [GMMP91], a high-level Petri net formalism for the description of time-dependent systems. The first tool to be included in SPADE is an SLANG interpreter. The developing effort is estimated in 2 man-years. The interpreter will be usable in two modes: simulation and run-time mode. In simulation mode, the interpreter will support design and evaluation of different process scenarios. In run-time mode, the interpreter will behave as a monitor of the environment, thus controlling all the events generated by humans and tool activations.

Accomplishments:
  The SLANG interpreter is still at a design level. The experimental design has begun on September 1992 and we hope to finish an experimental implementation by september 1993. The work will proceed in two directions: enhancing the SLANG interpreter and building new tools for SPADE. The expected enhancements include distributed execution and dynamic modification of the process, that is while it is being executed.

Published articles or reports:

[BFG92] S. Bandinelli, A. Fuggetta, C. Ghezzi, *Software Processes as Real Time Systems: A case study using High-Level Petri nets* to appear in the Proc. of International Phoenix conference on computers and Communications Arizona, April 1992.

[FG92] F. Ferrandina, S. Grigolli, *The SPADE project* CEFRIEL internal report, February, 1992.

[GMMP91] C. Ghezzi, D. Mandrioli, S. Morasca, M. Pezze, *A Unified High Level Petri Net Formalism for Time Critical Systems.* IEEE Transactions on Software Engineering, March 1991.

SPARK and the SPARK Examiner

Participants (project leader, group members):
  Bernard Carré, Jon Garnsworthy, Ian O'Neill, Dewi Daniels, William Marsh, Peter Amey plus many others in the past, notably Trevor Jennings, Fiona Maclennan and Paul Farrow.

Survey contact:
Bernard Carré,
Program Validation Limited,
26 Queen's Terrace,
Southampton SO1 1BQ,
U.K.
(tel. 0703-330001; fax 0703-230805)
(email: pvl@cix.compulink.co.uk)

Level of effort (person-years, duration):
Approximately 15 person-years on SPARK and the SPARK Examiner (since 1986); Approximately 6 person-years on proof support tools (since 1983).

Description:
SPARK is the SPADE Ada Kernel, an annotated subset of Ada. SPARK has relatively simple semantics in comparison with full Ada, giving a language suitable for formal reasoning, yet retaining considerable expressive power. Many of Ada's ambiguities and insecurities are eliminated in SPARK, and annotations (formal comments, ignored by an Ada compiler) can be used to specify the code's behaviour rigorously.

The SPARK Examiner checks conformance of a program text with the rules of SPARK, performing "flow analysis" to detect potential anomalies such as failures to initialise variables or unwanted dependencies between data. It can also be used to generate "path functions", which show the code's behaviour under different sets of input conditions. Finally, if the code is annotated with subprogram specifications (as pre- and postconditions), the Examiner can generate verification conditions for the proof of partial correctness of the code.

Verification conditions and/or path functions generated by the SPARK Examiner can be simplified by the SPARK Automatic Simplifier. This will attempt to eliminate non-traversible path functions, simplify the expressions which appear and prove the more straightforward verification conditions. The proof of any remaining verification conditions can be tackled interactively with the SPADE Proof Checker, which supports the proof attempt and generates a log of the proof steps carried out and proof rules used.

Accomplishments:

- The SPARK Examiner is written entirely in SPARK (apart from the body of the SPARK_IO package, which employs Ada TEXT_IO) and has been applied to itself (75,000 lines of code) to check conformance to the language rules.

- The SPARK Examiner is being used in the development of all the Risk Class 1 (safety-critical) and some of the Risk Class 2 (mission-critical) software for use in the European Fighter Aircraft (EFA).

- The SPADE Proof Checker has been used in the formal verification of modules of software for a jet engine fuel controller, and components of the Checker have been used in the proof of soundness of other of its components.

Published articles or reports:

- Bernard A. Carré, Jonathan R. Garnsworthy & William Marsh, *SPARK: A Safety-Related Ada Subset*, Proceedings of Ada UK 1992 Conference, October 1992 (to appear).

- Bernard A. Carré, Trevor J. Jennings, Fiona J. Maclennan, Paul F. Farrow and Jonathan R. Garnsworthy, *SPARK - The SPADE Ada Kernel*, (Edition 3.1), Program Validation Limited, 1992.

- Bernard A. Carré and Jonathan R. Garnsworthy, *SPARK - An Annotated Ada Subset for Safety-Critical Programming*, Proceedings of Tri-Ada Conference, Baltimore, A.C.M., December 1990.

- Bernard A. Carré, *Program Analysis and Verification*, in *High Integrity Software* edited by C.T. Sennett, pp. 176-197, Pitman, 1989.

- Bernard A. Carré, *Reliable Programming in Standard Languages*, in *High Integrity Software* edited by C.T. Sennett, pp. 102-121, Pitman, 1989.

Status:
All tools referred to are in existence and in use in industry.

Start date:
SPARK language: 1986.
SPARK Examiner: late 1987 (Version A), 1991 (Version B).
Proof Checker: 1983.

Completion:
The SPARK language was defined in 1988. It has undergone a number of fairly minor changes since then, following industrial experience of application of the tools and as part of the work on the formal semantics of SPARK.

The SPARK Examiner exists in two forms: Version A and Version B. Version A checks conformance to the language and performs flow analysis, and has been in industrial use since 1989. Version B additionally generates path functions and verification conditions, and was first released in 1991.

The SPARK Automatic Simplifier was released with Version B in 1991.

The SPADE Proof Checker was released to industry in 1988.

All tools are the subject of maintenance and upgrades in the light of industrial experience.

Future developments:
SPARK Examiner: a collection of viewing tools are under development, to allow the developer to navigate around a SPARK text and look at aspects of the code structure and design (e.g. calling hierarchy, cross-references).

Proof tools: an ongoing programme to enhance the proof tools exists, concentrating at present on the balance between automation and user-control and additional user-programmability.

Platforms: DECstation/Ultrix and HP are under consideration.

Strengths, weaknesses and suitability:

Strengths: All SPARK texts are also legal Ada, so any Ada compiler may be used. Provides a high-integrity subset of Ada, amenable to formal design and reasoning. Annotation language allows various Ada insecurities to be eliminated (e.g. aliasing through parameter-passing). The Examiner acts like a compiler, generating a listing file with warning and/or error messages embedded in the listing at appropriate points, making it easy to use.

Weaknesses: The current SPARK Examiner Version B has a restricted syntax for proof annotations, which needs to be extended. The Version B Examiner is not currently underpinned by a formal definition of SPARK (though this is being addressed).

External users:

Over 30; published users include GEC Avionics (Instrument Systems Division), Lucas Aerospace Limited and Eurofighter Jagdflugzeug GmbH and its subcontractors. Main applications areas of users include: aerospace; defence; signalling; automotive and other embedded software systems.

Applications:

- In the European Fighter Aircraft's more critical software.
- Airborne weapon management system for a military aircraft.
- In the development of the SPARK Examiner itself.
- The High Order Language Demonstrator (HOLD) programme, which investigated the use of formal specification methods and Ada in the development of a full-authority digital engine control system. (This was executed by Lucas Aerospace Limited for the U.K. Ministry of Defence.)
- Other embedded systems (details confidential).

Availability:

Licenses to use the tools are available from PVL. In Australia, sales and support are provided by Shayne Flint, of Rade Systems Pty Ltd (P.O. Box 136, Watsons Bay, NSW 2030; tel/fax: 02-2647165).

Current platforms are VAX/VMS, SUN-4 and 386-PC/MS-DOS.

Additional remarks:

Work on a formal semantics for SPARK is at an advanced stage; this work is being funded by the U.K. Defence Research Agency (DRA) at DRA, Malvern.

Other work on SPARK program design and verification issues is being undertaken by PVL. This includes a project funded by the U.K. Defence Research Agency (DRA), Farnborough, and a collaborative research project part-funded by the U.K. Department of Trade and Industry (Information Engineering Directorate), in which ICL Associated Services Division and the Universities of Cambridge and Kent are partners.

The flow analysis and verification capabilities of SPARK can also be applied to other source languages, via the related SPADE (Southampton Program Analysis and Development Environment) toolset, also supported by PVL.

TOPO

Participants (project leader, group members):
José A. Mañas,
Tomás de Miguel,
Tomás Robles,
Joaquín Salvachúa,
Gabriel Huecas,
Marcelino Veiga.

Dpt. Ingenieria de Sistemas Telematicos
Technical university of Madrid

Survey contact:
José A. Mañas,
Dpt. Ingenieria de Sistemas Telematicos
E.T.S.I. Telecomunicacion
Technical university of Madrid
E-28040 Madrid, Spain
Email: ¡lotos@dit.upm.es¿

Level of effort (person-years, duration):
about 30 man-years

Description:
Toolset to support product realization from LOTOS specifications.

It includes tools to perform semantics analysis, specification investigation (cross references, data type dependencies, etc.), and it is able to generate C or Ada code that may be compiled into a prototype.

It provides a compiler for data types, and a compiler for behaviour. These may be used independently. Usually data types are replaced by machine implementations (e.g. integers) to get a reasonable performance.

Performance is acceptable, even for final products, unless response times are very strict. Then, manual tuning is required.

Accomplishments:
Realistic specifications of up to 5000 LOTOS lines (commentless) have been successfully analysed and implemented on standard environments as UNIX (on sockets) and PC-MSDOS (using serial lines).

Published articles or reports:

Status:
Work on this environment is ongoing. A version is available and in the public domain for research and non-profit activities.

Start date:
October 1985

Completion:
Ongoing

Future developments:

We are currently working on extensions for

- distributed implementations
- test case generation
- test suite execution
- better integration with external data types (eg. ASN.1)

Strengths, weaknesses and suitability:

Strengths Able to handle big specifications, is robust and has been tested on many applications

Weaknesses It requires knowledge of LOTOS

Suitability It has been used for

- debugging
- validation
- rapid prototyping
- test suite execution
- real implementations

External users:
It is in anonymous FTP, so the number of people getting it is difficult to estimate. There may be a few hundreds of people that got it. It is harder to estimate the number of real users; there may be about 30 institutes and research departments.

Applications:

- rapid prototyping
- test suite execution
- real implementations

Availability:
It is in the public domain via anonymous ftp from
host: goya.dit.upm.es [138.4.2.2]
directory: info/src/lotos
New versions are made available as they get robust enough.

Alternatively, send two streamers to the contact address above; one is returned with the software.

Additional remarks:
No provision for commercial distribution, so far. Documentation could be really better.

TOPOSIM Performance Evaluation Tool

Participants (project leader, group members):
Department of Telematic System Engineering (DIT)
Polytechnical University of Madrid (UPM)

Survey contact:
Carlos Miguel Nieto
Dpto. de Ingenieria Telematica
E.T.S.I. Telecomunicacion
Ciudad Universitaria
E-28040 MADRID
SPAIN

tel: +34 1 5495700
tel: +34 1 5495762
ext: 437
fax: +34 1 5432077
E-mail: cmiguel@dit.upm.es
tlx: 47430 ETSIT E

Level of effort (person-years, duration):
Approximately 1.5 person years over one year. (this effort does not include the definition of the formal language on which the tool is based)

Description:
TOPOSIM is a performance evaluation tool based on an upward compatible LOTOS extension. It has been developed on top of an already existing full LOTOS compiler.

In the framework of the ESPRIT 5341 OSI95: "High performance OSI protocols with multimedia support on HSLAN's and B–ISDN", LOTOS has been extended in order to cope with the timing and probabilistic aspects of data communication systems. The extended language has been denoted as LOTOS–TP [3]. This formal language will permit us to obtain the performance figures of systems at early stages of design; e.g. mean transmission delay, the probability for a data packet to be lost in a network, and so on. Such a performance evaluation can be done by analysis or by simulation.

TOPOSIM obtains, by simulation, the performance figures of systems specified in LOTOS–TP. Since the evaluation is performed by simulation, the specifier has always to define the confidence level of the results that he desires to obtain. TOPOSIM runs the simulation until such a confidence level is obtained.

TOPOSIM has been developed on top of TOPO (LOTOS compiler developed under ESPRIT 2304 LOTOSPHERE project). It translates LOTOS-TP specifications into "annotated LOTOS" which can be compiled by TOPO and includes some additional external block to control the simulation.

Accomplishments:
TOPOSIM is based on an upward compatible LOTOS extension which has been formally defined. TOPOSIM has been used to evaluate the perfor-

mance figures of many examples, and also of real systems such as CODE (Cooperative Olympus Data Experiment) satellite data communication system [5], ISDN Q931 protocol, and MAC FDDI.

Published articles or reports:

1 C. Miguel, A. Fernandez, J.M. Ortuno and L. Vidaller, *A LOTOS based Performance Evaluation Tool*, To be published (accepted) in the forthcoming special issue of "Computer Networks and ISDN Systems" on: TOOLS FOR FDTs, 1992.

2 C. Miguel, A. Fernandez and L. Vidaller, *LOTOS Extended with Probabilistic Behaviors*, To be published (accepted) in "Formal Aspects of Computing. The international journal of Formal Methods", 1992.

3 C. Miguel, *Extended LOTOS Definition*, OSI95 ESPRIT II Project, OSI95/DIT/B5/8/TR/R/V1, January 1992.

4 C. Miguel, A. Fernandez, L. Vidaller, *Extending LOTOS toward Performance Evaluation*, (Submitted to) 12th International Symposium on Protocol Specification, Testing and Verification, Florida, USA, June, 1992.

5 A. Fernandez, C. Miguel and L. Vidaller, *Development of Satellite Communication Networks based on LOTOS*, (Submitted to) 12th International Symposium on Protocol Specification, Testing and Verification, Florida, USA, June, 1992.

6 C. Miguel, *Tecnicas de descripcion formal aplicadas a la evaluacion de prestaciones de sistemas de comunicacion*, PhD. Thesis, ETSI. Telecomunicacion, Universidad Politecnica de Madrid, March 1991.

Status:

Prototype tool completed in December 1991. It will soon be available for preliminary, controlled use. The application of TOPOSIM to real systems evaluations is ongoing.

Start date:

January 1991.

Completion:

December 1991

Future developments:

Improvements of the simulation control procedures. Depending on the foundings, improvements of the user interface, and tool efficiency. Support of every LOTOS extension defined in LOTOS-TP.

Strengths, weaknesses and suitability:

Strengths It inherits the strengths of TOPO tool on which it is based. It can be easily integrated in the LOTOSPHERE LOTOS environment (lite). It is based on an upward compatible LOTOS extension which is well defined and has a high expressive power.

Weaknesses The existing prototype tool consumes much more system resources (cpu time and memory) than classical, non-formal simulation tools.

Suitability It is suitable to evaluate performance figures of systems specified formally. So, LOTOS based specifications can be used to support completely the design (including performance evaluation) of concurrent systems.

External users:
None

Applications:
Performance evaluation of systems. Evaluation of systems properties in terms of their probability of verification (not only in terms of must or may).

Availability:
TOPOSIM is available as a prototype tool. It runs on SUN workstations.

Additional remarks:
None

VEDA 2.0

Participants (project leader, group members):

Verilog Bernard Algayres, Jean Michel Ayache, Veronique Coelho, Laurent Doldi, Hubert Garavel, Yves Lejeune, Daniel Pilaud, Carlos Rodriguez.

CNET Roland Groz, Jean Francois Monin.

LGI/IMAG Joseph Sifakis.

University of Rennes Claude Jard, Thierry Jeron.

Survey contact:
Bernard Algayres
Verilog S.A.
150 Rue Vauquelin
31081 Toulouse - France
tel: (33) 61 19 29 39
email: verilog@verilog.fr

Level of effort (person-years, duration):
Approximately 12 man years.

Description:
In VEDA, descriptions are made in ESTELLE, a structured language based on extended finite state machines and PASCAL data types. ESTELLE is an ISO international standard since 1989.

In addition to traditional verification and simulation capabilities, VEDA provides features for *automatic analysis*:

- deadlock and loops detection,
- formal verification of properties,
- automatic validation of expected behavior,

- generation of scenarios for testing.

The VEDA simulator provides for both *interactive and intensive simulation*. During interactive simulation, a complete set of commands is available to play scenarios and to observe parts of the model such as variables, queues and other state related information. Intensive simulation is based on random execution of the specification. The simulator interface is based on a menu driven multi-windows interface running under X windows/Motif.

The VEDA verifier provides *formal proof* of the specification and *scenario generation* by means of reachability analysis. Formal properties of the specification or test objectives must be specified by the user with the observation language, a variant of ESTELLE. The verifier is designed to explore all the execution paths and dynamic states of a specification, while checking in each state that properties are verified or tests objectives fulfilled.

The VEDA 2.0 verifier has been run on specifications producing over a million states in less than a couple of hours on conventional SPARC machines.

Accomplishments:
Used on over 10 industrial protocol systems in various sectors (aerospace, defense, telecommunications and transport).

Published articles or reports:

- B. Algayres, V. Coelho, L. Doldi, H. Garavel, Y. Lejeune and C. Rodriguez, *VESAR: Un outil pour la specification et la verification formelle de protocoles*, Proceedings of the CFIP 91 (Colloque Francophone sur l'Ingenierie des Protocoles, Pau, France, September 1991.

- B. Algayres, V. Coelho, L. Doldi, H. Garavel, Y. Lejeune and C. Rodriguez, *VESAR: A pragmatic approach to formal specification and verification*, to appear in Computer Networks and ISDN networks, Special Issue on Protocol Specification and Verification, 1992.

- B. Algayres, V. Coelho, L. Doldi, H. Garavel, Y. Lejeune and C. Rodriguez, *VEDA 2.0: Formal specification and verification of communication systems*, Verilog Internal Report, 1992.

Status:
Commercialized Product. 40 licenses sold.

Start date:
June 1989.

Completion:
June 1991 (as a commercial product).

Future developments:
Introduction of the following:

- Full graphic Editor

- TTCN tests suite production

- Performance simulation

Strengths, weaknesses and suitability:

Strengths Easy to learn. The fastest reachability graph analysis to date. Various strategies for carrying verification.

Weaknesses Non-graphic interface. Estelle for data representation (i.e. Pascal).

External users:
EDF, Thomson, Setics, Cap Sesa, CNET, Alcatel, Elecma, Universities of Montreal, Paris, Grenoble, Toulouse, Rennes and various schools.

Applications:
Communication systems specification and validation.

Availability:
Commercial Product.

Additional remarks:
VEDA has been developed in joint effort with CNET (French Telecom R&D center) and Universities of Grenoble and Rennes.

Veritas

Participants (project leader, group members):
Keith Hanna, Neil Daeche, Gareth Howells, Mark Longley

Survey contact:
Keith Hanna
Electronic Engineering Laboratories
University of Kent
Canterbury,
Kent CT2 7NT
United Kingdom
e-mail: fkh@ukc.ac.uk

Level of effort (person-years, duration):
12 man-years

Description:
Veritas is a design logic (that is, a formal logic intended for describing and reasoning about designs) intended for formal verification of digital hardware and low-level software. It is a higher-order logic (classical) that incorporates dependent types and subtypes. It is particularly effective at describing the bounded, parametrized types typically found in hardware, such as:- bit, n-bit word, fixed length numeral to base n, etc.

It has been used for Formal Synthesis, an approach to design in which the activities of behavioural synthesis and of formal verification are combined: the starting point is a behavioural specification, the end result is an implementation together with a proof of its correctness.

The potential disadvantage of using dependent types (rather than polymorphic types) is that type checking is undecidable. In practice, this difficulty can be almost entirely overcome by suitably designed tactics working within a goal-directed framework.

The Veritas logic has been implemented both in Standard ML and in Haskell. Both implementations provide an X-windows user interface that allow proofs to be constructed largely by point and click operations. The Veritas-90 system is freely available for evaluation by academics.

Accomplishments:

Published articles or reports:

- F K Hanna, N Daeche, M Longley, *Specification and Verification using Dependent Types* pp949-964, Trans IEEE on Software Eng, Vol 16, No 9, Sept 1990.

- F K Hanna, M Longley, N Daeche, *Formal Synthesis of Digital Systems*, pp153-169, in "Formal VLSI Specification and Synthesis (I)", edited L J M Claesen, Elsevier, IFIP, 1990

- F K Hanna, N Daeche, *Dependent Types and Formal Synthesis*, to appear in Phil Trans of the Royal Society (A), April 1992.

- F K Hanna, N Daeche, G Howells, *Guide to the Veritas Design Logic*, Technical Report, 1992, University of Kent.

Delft VDM-SL Front-end

Participants (project leader, group members):
Delft University of Technology

Survey contact:
Nico Plat
Delft University of Technology
Faculty of Technical Mathematics and Informatics
P.O. Box 356, NL-2600 AJ Delft, The Netherlands
Phone +31-15784433
Fax +31-15-787141
E-mail nico@dutiaa.tudelft.nl

Level of effort (person-years, duration):

Description:
The Delft VDM-SL front-end has the following functionality:

- Syntax checking. The front-end accepts the ASCII syntax for the BSI/VDM-SL, the notation currently being standardized by ISO/IEC JTC1/SC22/WG19 employed by the formal method VDM. Simple error recovery facilities are provided.

- Static-semantics checking. The major part of the static-semantic checking is type checking. The type system that has been implemented is described in [plat90]. Other checks include scope checking and various minor checks.

- Generating an intermediate representation of the specification. The front-end generates an abstract form of the original specification, which is called 'intermediate representation' because it is intended to be used by other tools, e.g. a pretty-printer or a prototyping tool. The form of the intermediate representation itself can be easily maintained by an in-house developed tool called DIAS.

The front-end is batch-oriented, so, as is the case with conventional compilers, the user must provide an input file with an ASCII VDM specification, which is then analyzed by the front-end.

The user has the following options:

- having only error messages (if any) generated by the front-end;

- having a listing of the input generated, annotated with error messages;

- using the front-end in combination with an editor, so that erroneous specifications can directly be changed, after which a new check-edit session can directly be started.

The implementation of the front-end is based on an attribute grammar (AG). AGs were introduced by Knuth as a method for defining semantics of programming languages. An AG is based on a context-free grammar, in which both the terminal and nonterminal symbols can be augmented with attributes. An attribute is used to hold a semantical value associated with a symbol in the grammar. By defining relationships between attributes, semantic properties of the programming language can be defined.

A system capable of evaluating the attribute values for a specific programme (or in our case: a VDM specification) is called an attribute evaluator. For most classes of AGs such attribute evaluators can be automatically generated. We have used the GAG system for the implementation of the front-end.

Accomplishments:

Published articles or reports:

- Nico Plat, *Towards a VDM-SL Compiler*, Delft University of Technology, December 1988

- Nico Plat, Ronald Huijsman, Jan van Katwijk, Gertjan van Oosten, Kees Pronk, Hans Toetenel, *Type checking BSI/VDM-SL*, in VDM and Z; Proc. of the 3rd VDM-Europe Symposium, Springer-Verlag, Lecture Notes in Computer Science 428 1990

- Nico Plat, Kees Pronk, Marcel Verhoef, *The Delft VDM-SL front-end*, Formal Software Development Methods; Proc. of the 4th VDM-Europe Symposium, Springer-Verlag, Lecture Notes in Computer Science 551, 1991

Status:

The front-end has been implemented as described, and is currently being tested.

Start date:
  July 1988

Completion:

Future developments:

- Implementation of various enhancements, such as prototyping facilities, pretty-printing facilites, support of LaTeX formatted specifications, cross-reference generating, etc.

- Integration with an already existing proof assistent (Mural from Manchester University) as part of the ESPRIT-III project AFRODITE.

- An extension to support an object oriented version of VDM, called VDM++ as part of the ESPRIT-III project AFRODITE.

Strengths, weaknesses and suitability:

  Strengths efficient, easy to use.

  Weaknesses has not yet been extensively tested.

External users:
  CAP Gemini Innovations, GAK Amsterdam, RU Utrecht, The Netherlands.

Applications:

Availability:
  Upon request

Additional remarks:
  None.

B   Survey of Formal Methods Applications

Simplified Reliable Transfer Service

Participants (project leader, group members):
Tomas de Miguel, Tomas Robles, Joaquin Salvachua, Arturo Azcorra

Survey contact:
Arturo Azcorra,
Dpto. Telematica,
E.T.S.I. Telecomunicacion,
Ciudad Universitaria s/n,
E-28040 MADRID,
SPAIN

Tel: +34 1 5495700 ext: 375, 368
telefax: +34 1 5432077
email: aazcorra@dit.upm.es
telex: 47430 ETSIT E

Level of effort (person-years, duration):
One man year

Description:
The project was a study on the results achieved by using a LOTOS-based design methodology. The study was performed taking a simple protocol — Simplified RTS— as the system to deal with. The work done was the service requirements capture, protocol design, protocol implementation and finally the maintenance of the system during two years.

The protocol entity had to be integrated in a larger system –an X.400 message handling system and also integrate existing software.

A relevant feature of the experience is that the translation from the LOTOS specification to C language was done both using the compiler TOPO and also manually. Another relevant feature is that the maintenance phase has been covered in the case study.

Accomplishments:

Published articles or reports:
T. de Miguel, T. Robles, J. Salvachua, A. Azcorra. "The SRTS experience: Using TOPO for LOTOS Design and Realization". Procs. of Formal Description Techniques, November 1990, Madrid, SPAIN.

Current Endorsed Tools List Example (CETLE); contract for DoD by Loral Command & Control Systems

Participants (project leader, group members):
Leon Buczkowski, Steve Eckmann, Jim Freeman, Richard Neely

Survey contact:
James W. Freeman,

freeman@cos1.cos.loral.com;
4220 Bromley Place,
Colorado Springs,
Colorado, 80906;

Government contact for Loral delivered documentation:
Karen Ambrosi,
Department of Defense;
kambrosi@dockmaster.ncsc.mil

Level of effort (person-years, duration):
2.3 man-years

Description:
Loral Command & Control Systems (Loral) has produced a technology demonstration for the government designed to meet the A1 level of assurance as specified by the Trusted Computer System Evaluation Criteria (TCSEC), DoD 5200.28-STD. The technology demonstration includes the design of an example system that illustrates the state of the art in the application of formal design methods to a small development project at the A1 level of assurance. The objectives of the project were (a) to demonstrate the state of the art in security assurance technology, (b) to get new people involved in the practical application of this technology, and (c) to make the resulting information available to as wide an audience as possible.

A significant amount of the effort was accomplished by team members who were not familiar with Gypsy and the Gypsy Verification Environment (GVE) nor with the particular security modeling approach taken that utilized Gypsy and the GVE.

The team defined the example system, termed the DataFix Sysem, as one having a small number of commands that would allow a user to create and delete files as well as allow the communication of authorized information to other users in a MLS environment, all in support of a potentially larger command and control mission. The team then produced a system design that incorporated an explicitly defined Trusted Computing Base (TCB). The design allows for various application processes to be executed under TCB control, which includes a file system. A terminal manager that incorporates a trusted shell and an application manager completes the top-level design of the TCB, the software portion of which is called MicroSys. MicroSys is supported by an underlying microkernel that provides for process isolation and inter-process communication.

The approach taken to generate the necessary formal specification design structure and related security conditions focused on expressing system behavior at interfaces. Specifically, one expresses the external behavior of a given component's inteface, and then relates that behavior to the external behavior of other components. Such a component could be the entire TCB or a component within the TCB, say the Terminal Manager. In this way, lower level properties can be linked to higher level properties in a modular way so as to establish the desired result of the consistency of the formal design specification with the formal model of the security policy.

The particular way by which the team used Gypsy to express the system structure and design is termed the "relational form". It allows one to produce a specification having the focus on capturing the external behavior of components. The approach builds on and extends previous work that compares and contrasts the relational form to both a procedural form and a functional form of producing a formal specification.

Lessons learnt: see Final Report for full discussion;

- Formal specification and verification technology can be learned and it can be applied in an effective manner to produce a product.

- The process is not reduced to practice as compared to the overall system and software development process. Yet the formal process is defined as well as the system development process. There can be a significant amount of up-front effort in an A1 development.

- There is value in accomplishing a covert channel analysis on a system definition and top-level system design prior to producing the full formal specification and proofs.

- There is value in applying formal methods within a development process, yet that value needs to be assessed in light of the cost-benefit of the overall development process.

- There is value in having an explicit system specific security policy, rather than depend on more general policty statements that occur in the DoD literature, including the policy statements within the TCSEC.

- A system can be developed using Gypsy and GVE in a reasonably effective manner. To achieve that effectiveness, however, requires a careful consideration of several issues that is to result in a cohesive approach. The performance of the GVE can be enhanced considerably by how one approaches the proof process.

Accomplishments:

Published articles or reports:

- *Achieving Understandable Results in a Formal Design Verification,* R.B. Neely, J.W. Freeman, M.D. Krenzin, Proceedings of The Computer Security Foundations Workshop II, June 11-14, 1989, pp. 115-124.

- Invited participation on forthcoming panel on Verification at 1992 National Computer Security Conference (NIST / NSA)

Contract documentation available through K. Ambrosi:

- Requirements Report
- System Security Definition: System Security Policy and Formal Model
- System Definition and Modeling Approach
- Descriptive Specification

- Formal Specification
- Verification Demonstrations
- Covert Channel Analysis
- Final Technical Report (with lessons learned and tracking of effort expended)

Data Handling Subsystems for a VSAT Satellite Network (CODE)

Participants (project leader, group members):
Department of Telematics Engineering, Polytechnical University of Madrid, Spain. Telefonica Sistemas Spain. Teice Control Spain.

Survey contact:
Angel Fernandez
Associate Professor Dpt. Ing.Sist.Telematicos
E.T.S.I. Telecomunicacion
Ciudad Universitaria
MADRID (SPAIN)
E-28040

E-mail: afernandez@dit.upm.es
tel: +34 1 5495700 x.437
fax: +34 1 5432077
tlx: 47430 ETSIT E

Level of effort (person-years, duration):
27 man months

Description:
CODE Network is an ESA(European Space Agency) funding VSAT satellite network based on Olympus Satellite. The network is experimental in two senses. First because of the radio frequency bands it uses (Ka band). Second because of the satellite link speeds (4Mbps broadcast, 128Kbps from terminals).

From the formal methods point of view, the interest of this project is to demonstrate the possibility of having quite separate design and codification teams. In this way the Department of Telematics Engineering designed the protocols for the network giving as a final product a LOTOS specification of the whole system. This specification was taken as input by Telefonica Sistemas to make the actual codification of the system.

Within this development LOTOS compiler (TOPO) and LOTOS simulator HIPPO were used. Through these tools the test cases designed for the whole system were run against the actual LOTOS specification of the system giving further confidence in the specifications. These tests were also coded by Telefonica sistemas being the basis for the validation phase between the specification and the coding. Further tests, mainly white box, were produced to assert the specific design decisions and physical limitations imposed in the coding phase.

The experience produces quite a few, about twelve, interactions between the specification and the coding groups. But it also showed that it is not enough to produce abstract specifications of the system and pass them to the coding team. It is necessary to transform such abstract specifications in two ways: first to take into account the performance characteristics of the system being described, second to take into account the specific hardware configuration where the actual code is going to run.

As an example of the first transformations, a layer-wise abstract specification of the system protocol stack was transformed into a new one were data packets were handled quite differently from most of the signaling packets because of their different time requirements.

As an example of the second transformations, the efficient specifications taken from the transformations above were modified again to take into account the different hardware architectures of the terminals and the central station of the network. Since the performance and storage characteristics, the terminals were single CPU interrupt driven systems, and the central station was a common bus multiprocessor system. Facts, such as interprocessor communication with their corresponding dead-live lock problems, were taken into account for the final specification of the central station.

Accomplishments:

Published articles or reports:

- *Satellite Link Protocols Design for the CODE system.* A Fernandez, L.Vidaller, C.Miguel, D.Briones. Proceedings of the Olympus Utilisation Conference. Vienna 12-14 April 1989. ESA SP-292, May 1989.

- *Development of Satellite Communication Networks Based on LOTOS.* A. Fernandez, C.Miguel, L.Vidaller. Submitted to the 12th International Symposium on Protocol Specification, Testing, and Verification.

Hewlett Packard Leadership Projects.

Participants (project leader, group members):
HP Labs Software Engineering Department and various product development engineers.

Survey contact:
Dr Steve Bear
Hewlett Packard Laboratories Bristol
Filton Road
BRISTOL UK
BS12 6QZ

phone: +44 272 799910
fax: +44 272 228003
e-mail: sb@hplb.hpl.hp.com
X.400: G=steve; S=bear; OU1=hpl; OU2=unix; O=hp; P=hp; A=gold 400; C=gb

Level of effort (person-years, duration):
Approximately 8 person-years since 1989

Description:

Background:

The HP Labs Software Engineering Dept is working with product development teams within HP to transfer and to apply formal specification technologies within the industrial environment of HP. In a series of "leadership" projects, product engineers have been trained to use the formal specification language HP-SL, and have developed formal specifications of product behaviour in HP-SL. Most of the products developed have been medical devices with embedded software. The engineers involved have been experienced sofware developers, but without any prior experience of formal techniques.

Role of formal methods in the development process:

In each project, the use of formal techniques has focused on developing a clear early, description of the intended behaviour of the product. The resulting formal specification documents have been used as a firm foundation for the remainder of the development. In a few of the projects, design descriptions have also been partially formalised. There have been no attempts to prove an implementation relation.

Benefits:

The main benefit of using formal specification in these projects has been to improve the quality of early decisions about the required behaviour of the product. The result has been to reduce the time required to get the product to market.

The benefit has arisen in a number of ways:

- The definition of product behaviour has been produced earlier, has been more precise and more complete.

- The definition of product behaviour has been easy to review.

- The definition of product behaviour has provided clear guidance for later stages of the development.

The formal techniques have been used to assist the *social* process of software development. The overall impact has been to reduce design and implementation rework. This has reduced the time required to develop the product.

Drawbacks:

Engineers using formal specification for the first time face a significant learning curve. They have to cope with a new language and a new way of thinking about systems. Post training consultancy and support is necessary to move from text book examples to industrial scale applications.

Lessons and recommendations:

The projects have provided two important lessons about the role of formal methods in industry.

Firstly, given appropriate training and support, non-specialist engineers can use formal specification techniques in an effective way. Post-graduate training in mathematics or computer science is not necessary.

Secondly, formal specification techniques improve the efficiency of the software development. Rework is significantly reduced, and overall development time is shortened. Formal techniques are not limited to critical products.

The experience allows us to recommend that industrial exploitation of formal methods for non-critical products should concentrate on using formal specification to improve the quality of early decisions about product behaviour.

Accomplishments:

Published articles or reports:

- S.P. Bear and T.W. Rush, *Rigorous Software Engineering: A Method for Preventing Software Defects*, Hewlett Packard Journal, Dec 1991, 24-31.

- P.C. Goldsack and T.W. Rush, *Specifying an Electronic Mail System with HP-SL*, Hewlett Packard Journal, Dec 1991, 32-39.

- P.D. Harry and T.W. Rush, *Specifying Real Time Behaviour in HP-SL*, Hewlett Packard Journal, Dec 1991, 40-45.

- B.R. Ladeau and C.W. Freeman, *Using Formal Specification for Product Development*, Hewlett Packard Journal, Dec 1991, 46-50.

- J.L. Cyrus, J.D. Bledsoe and P.D. Harry, *Formal Specification and Structured Design in Software Development*, Hewlett Packard Journal, Dec 1991, 51-58.

Modular Security Device (MSD)

Participants (project leader, group members):
Application developer:
Naval Research Laboratory, Washington DC, Eather Chapman, Miyi Chung, David Kim, Ken Hayman (Australian DSTO), Richard Hale, David Mihelcic, Andrew Moore, Charles Payne, Maria Voreh

Hardware developer:
E-Systems, Inc., Texas

Survey contact:
System Engineer :
David Mihelcic;
mihelcic@itd.nrl.navy.mil
202-767-73411

Software Engineer:
Andrew Moore
moore@itd.nrl.navy.mil

202-767-6698

both at: Code 5540
Naval Research Laboratory
Washington, DC 20375-5000

Level of effort (person-years, duration):
7 person years so far; 2 years duration

Description:
The Modular Security Device Family is a class of systems intended to provide data security for communications in a networked environment. The primary security policy for the networked environment is data confidentiality. The primary means of enforcing this policy in the network is cryptography; each MSD family member supports the cryptographic control function for the network. The network is responsible for ensuring the separation of sensitive (unencrypted) data from non-sensitive (encrypted or header) data and for ensuring that all communications between users through the network are permitted by a pre-defined connection plan. All members of the MSD Family must satisfy the following critical properties:

- The sensitive portions of incoming user messages must be encrypted.

- Each communication must conform to the predefined connection plan.

- The rate at which plaintext information may bypass the cryptographic function must be limited.

- The format of the data bypassing the cryptographic function must satisfy pre-defined criteria.

The current effort involves developing the cryptographic control software and hardware for one member of the MSD Family, called the ECA. A goal of this project is to define a development methodology that integrates informal (intuitive) descriptions and arguments with formal specifications and proofs in a coherent manner. The basic components of our baseline development methodology include

- NRL's Software Cost Reduction (SCR) Software Development and Documentation Methodology;

- i-Logix's Statemate (tm) computer-aided system engineering tool (with a graphical simulation capability) founded on Harel's statechart formalism;

- Hoare's Communicating Sequential Processes (CSP) specification language and Trace Model; and

- ORA (Canada) Corporation's mEVES verification system.

We hope to refine the methodology through its application to the ECA so that, through increased formality and reviewability, we gain high assurance that critical properties are satisfied.

What role did the formal methods play in the development process? Formal methods are closely integrated with the development process, from critical requirements modeling to coding. Functional requirements are documented in a rigorous, yet intuitive, tabular format, using techniques that recently evolved out of the SCR work. The critical (security) requirements are specified using the Trace Model of CSP, so as to allow formal specification (and proof) of the ECA as a network of communicating components. Statemate is used to animate the functional requirements and design the top level architecture of the ECA. This architecture is reflected in a CSP description; a formal decomposition of the critical requirements onto the major components of the architecture is performed using the CSP proof theory and a method developed at NRL. This decomposition is performed down to the level of sequential CSP processes. Once the major programs are defined using the modular decomposition approach of SCR, the trace specifications are translated into pre and post conditions on mVerdi (the language of mEves) programs. Each program is then implemented and verified to satisfy its specification, and finally translated to an m-Verdi-like subset of Ada. At the time of writing, we have identified 30 modules for implementing the ECA system. Where practical we have declared objects used for specification in the same module as the executable programs they specify. In some cases, it is more practical to define specification objects in separate modules, for example, as it would be for a generally useful theory of sequences. Of the 30 modules identified (of which 10 are used for specification purposes only), 16 of the 20 implementation modules identified have been translated to Ada.

What were the benefits? We may find some benefits and drawbacks to our methodology during integration testing, external review and actual fielding of the device that have not arisen at our current stage of development. The primary benefits perceived at this time include

1. The formal proof process uncovered several security-critical errors that might not have been found otherwise. Although it is difficult to know for sure, we expect that many of these problems could have been found by conventional review and testing methods. It is difficult to compare the cost of formal methods with the gain achieved from catching errors earlier in the process.

2. The use of formal specification early in the design process required much more thinking up front than we probably would have done without them. We believe this led to a design and implementation with fewer flaws (both security-critical and otherwise).

3. The use of formal methods has increased the development team's confidence that the system will satisfy specified critical properties. The mere process of formalization has strengthened our understanding of the system and its critical requirements. This increased understanding has led to improved performance and job satisfaction among the team members.

4. Because it supports proofs of programs, automatic reconstruction of proofs, and automatic application of rewrite rules, mEves appears to have been a wise choice for the ECA development. These features are necessary in moving target developments; they reduce the amount of human intervention required and enhance proof automation.

What were the drawbacks? The primary drawbacks we have perceived at this point in time include

1. Difficulty in training personnel to use the techniques. The large learning curve associated with the techniques caused a formal methods bottleneck in the development process. This is partly due to the relatively immature state of the tools and techniques used. We need courses that show how to use the techniques effectively. Two of the techniques we used, CSP and the decomposition methodology, have no associated mechanical tools. It is clear that tool support for verifying concurrent systems would be advantageous.

2. Proofs produced with mEves are difficult to review; they provide either too little information or too much. Human review is very important for developing high assurance systems using tools that have not been verified themselves. To get the most out of formal methods, the developer needs to convince others of the correctness/trustworthiness of an application.

3. The Ada programmers felt overly constrained by the strict standard we developed for manually translating mVerdi programs to Ada. This drawback would be eliminated were automatic translators produced, or better yet, increased support for directly executing the verifiable language.

What lessons can be drawn, recommendations made?

1. For building non-trivial systems, more support is needed in the area of proof management. It should be possible to change a portion of a previously proven module without having to redo all the proofs of programs that use that module. It should be possible to mechanically check for only those programs that depend on the modified portion to limit the effects of modifications. We found that the SCR concept of information hiding was the only (manual) way we had to limit the effects of changes.

2. Tools for verifying properties about concurrent systems are needed. Even a parser and graphical simulation tool for a powerful process algebra would be helpful. Methods for hierarchical communication event refinement, e.g., in CSP, are needed to make proofs of complex systems more practical.

3. The availability of a simple proof checker that has been subjected to rigorous analysis would increase the acceptability of basing assurance arguments on the results of heuristically driven theorem provers. This would also reduce (although not eliminate) the need for human review.

Accomplishments:

Published articles or reports:

- Moore, A.P., *The Specification and Verified Decomposition of System Requirements using CSP*, IEEE Transactions on Software Engineering, Vol. 16, No. 9, Sep. 1990.

- *Modular Security Device Family Software Engineering Methodology,*
  NRL Technical Memorandum 5540-017:AM:am.

- *The ECA Critical Requirements Model,* NRL Technical Memorandum
  5540-097:CP:cp.

Data Handling Subsystem for the Mobiles of a Mobile Satellite Network (PRODAT)

Participants (project leader, group members):
Department of Telematics Engineering, Polytechnical University of Madrid.

Survey contact:
Angel Fernandez
Associate Professor
Dpt. Ing.Sist.Telematicos
E.T.S.I. Telecomunicacion
Ciudad Universitaria
MADRID (SPAIN)
E-28040
E-mail: afernandez@dit.upm.es

tel: +34 1 5495700 x.437
fax: +34 1 5432077
tlx: 47430 ETSIT E

Level of effort (person-years, duration):
Five man years.

Description:
PRODAT is a mobile satellite network developed under ESA (European Space Agency) contracts. The development included several phases and the phase corresponding to this project tried to demonstrate the feasibility of ESA designed link layer protocols and access methods. The demonstration consisted of the prototyping of several mobile terminals (terrestrial, maritime, and aeronautical) and a central station. And of field experimentation using the MARECS satellite. Several partners manufactured differents prototypes, including DORNIER (Germany), RACAL (UK), SNEC (France), Spain being in charge of providing (3) aeronautical and (2) maritime terminals.

From the point of view of formal methods, the project assessed the feasibility of using FDTs (Formal Description Techniques) for the design and development of protocol architectures on real time embedded multiprocessor systems.

The project lasted since January 1986 until November 1987. For the time being no tools for FDT design exist other than limited syntax checkers. So, a manual methodology was established to translate the FDT (LOTOS) used during the design phase of the project to Modula-2 language used for the actual coding. The methodology was based on an ad-hoc kernel providing LOTOS process and synchronisation facilities.

The benefits obtained from the use of FTDs were twofold. The first relates to the availability of formal texts covering the whole system from the very beginning steps of the Software Life Cycle; including not only the interfaces between different parts but their whole semantics, allowed to analyse and modify such designs as they were being understood. Through this analysis sophisticated modules were reduced up to one tenth of their initial size, in lines of text, without writing a single line of code.

The second relates to the facility of producing actual code from a formal text describing the system instead of from natural language descriptions.

The drawbacks that emerged from the experience concerned the huge amount of actual lines of code produced by the methodology used to translate FDTs lines of text into actual code, and the difficulty in going deeply enough with the FDTs to describe hardware subsystems such as interrupts, DMA channels, etc. This difficulty seems to preclude the fully automated translation from FDTs into actual language.

Another major drawback of the project was the difficulty of designing and implementing using abstract data types. This difficulty was so great that the data types used in the specification included only their signatures (interfaces)

For the rest of the FDT LOTOS, junior engineers took 1 month to be able to read the specifications written in LOTOS and 3 months to write their own.

Accomplishments:

Published articles or reports:

- *PRODAT - The derivation of an Implementation from its LOTOS Formal Specification.* A. Fernandez, J. Quemada, L. Vidaller, C. Miguel. Protocol Specification testing and Verification VIII, Elsevier(North Holland) 1988.

- *LOTOS Based Derivation Methodology in the PRODAT Project.* A. Fernandez, J. Quemada, L. Vidaller, C. Miguel. IFAC Real Time Programming, Valencia, Spain 1988.

Formal Definition of Turing Language

Participants (project leader, group members):
R.C. Holt, J.R. Cordy, J.A. Rosselet, P.A. Matthews

Survey contact:
R.C. Holt
holt@csri.toronto.edu
Dept of C.S.
Univ of Toronto
Toronto M5S 1A4 CANADA

Level of effort (person-years, duration):
3 person yr.

Description:

- Essentially all aspects of a production programming language were formally specified. The formal language used for the specification was Rosselet's ADL (axiomatic denotational language). See "The Turing Programming Language: Design and Definition" [Prentice-Hall 1988, pg. 325] for details.

- The formal definition served as the basis for the "turbo" implementation of Turing, which compiles at 60,000 lines/min on a Sun/4. Forty percent of Ontario highschools use this commercial product on microcomputers.

- The formal definition clearly laid out a very complicated entity (a programming language). It made subsequent implementation more rapid and reliable.

- It was harder to write the formal definition than it was to write the compiler. The formal definition is not accessible to anyone but a Computer Science expert.

- What lessons can be drawn, recommendations made? Formalism in the large is extremely challenging. Much of the potential benefit of using the formalism can be lost if the result is not easily readable by the intended audience. Even simple tools, eg, syntax checkers, are very imporant. Prototyping by executable specifications are a very big help in validating large specifications. Any large information entity, such as a language specification, almost inevitably has errors in it.

Accomplishments:

Published articles or reports:
See reference to book above. See also: *The Turing Programming Language*, Holt and Cordy, Dec 88, CACM.

Using VDM to Specify OSI Managed Objects

Participants (project leader, group members):
Lynn Marshall and Linda Simon

Survey contact:
Lynn Marshall
Computing Research Laboratory
Dept 0R00 GTWY
Bell-Northern Research Ltd.
P. O. Box 3511 Station C
Ottawa, Ontario
Canada K1Y 4H7

ph: (613) 765-4856
fax: (613) 763-4222
e-mail: lynnmar@bnr.ca

Level of effort (person-years, duration):
several person-months

Description:

Outline the nature of the application, the tools/techniques employed. We used an Object-oriented flavour of VDM to formalize the standards describing OSI Managed Object behaviour.

What role did the formal methods play in the development process? Currently the standards use ASN.1 to describe the data structures, but the behaviour is described in English. Formalizing the standard was the goal of our work.

What were the benefits? Ambiguities in the standard were uncovered during our work.

What were the drawbacks? Education of the community is needed before formal standards will be accepted.

What lessons can be drawn, recommendations made? Formal standards are needed. We will recommend our approach to the standards bodies.

Accomplishments:

Published articles or reports:

- Linda Simon and Lynn S. Marshall *Using VDM to Specify OSI Managed Objects* Proceedings IFIP TC/WG6.1, 4th International Conference on Formal Description Techniques (FORTE'91), Sydney, Australia, Nov. 1991, Eds. G. A. Rose and K. R. Parker, Pub: Elsevier.

- Lynn S. Marshall and Linda Simon. *Using VDM within an Object-Oriented Framework*, Proceedings VDM'91, Formal Software Development Techniques, LNCS 551, October 1991.

C   Acronyms and Trademarks

Acronyms

AMN Abstract machine Notation

CAA Civil Aviation Authority

CCS Calculus of Communicating Systems

CESG Communications Electronic Security Group (UK)

CICS Customer Information Control System

CLI Computational Logic Inc (US)

CSE Computer Security Establishment (Canada)

CSP Communicating Sequential Processes

DARPA Defence Advanced Research Projects Agency (US)

DFD Data Flow Diagram

DND Department of National Defence (Canada)

DoD Department of Defence

DRA Defence Research Agency

FTLS Formal Top Level Specification

HOL Higher Order Logic

JSP Jackson Structured Programming

KLOC Thousand Lines Of Code

LCF Logic of Computable Functions

LOC Lines Of Code

MOD Ministry Of Defence

NASA National Aeronautics Space Administration (US)

NCSC National Computer Security Centre (US)

NSA National Security Agency

PICT Programme on Information and Communication Technologies

PRG Programming Research Group

QA Quality Assurance

QC Quality Control

VDM Vienna Development Method

VIPER Verifiable Integrated Processor for Enhanced Reliability

Trademarks

The following are trademarks owned by their respective organisations.

ADA

ELLA

Ina Jo

Lucid

MS-DOS

Oracle

OSF

POSIX

RAISE

RDD

RSL

Spade

Statemate

UNIX

VHDL

X/OPEN

List of Contributors

Jean-Raymond Abrial
26 Rue des Plantes, 75104 Paris, France
fax: 33 1 40 44 50 12

Chapter 4: On Constructing Large Software Systems
© Jean-Raymond Abrial 1993

Victor Basili
Department of Computer Science, University of Maryland at
College Park, College Park, MD 20742, USA
e-mail: basili@cs.umd.edu

Chapter 8: Software Quality: A Modelling and Measurement
View © Victor Basili 1993

Dan Craigen
Odyssey Research Associates Canada, Suite 506, 256 Carling
Avenue, Ottawa, Ontario K1S 2E1, Canada
e-mail:dan@ora.on.ca

Chapter 9: Modelling Working Group Summary © Dan Craigen
1993

Joseph Goguen
Oxford University Computing Laboratory Programming
Research Group, 11 Keble Road, Oxford OX1 3QD, UK
e-mail: Joseph.Goguen@prg.ox.ac.uk

Chapter 1: Introduction © Joseph Goguen 1993

Antony Hall
Praxis Systems plc, 20 Manvers Street, Bath BA1 1PX, UK
e-mail: jah@praxis.co.uk

Chapter 6: Integrating Methods in Practice © Praxis Systems plc
1993

Michael Jackson
101 Hamilton Terrace, London NW8 9QX, UK
e-mail: attmail!jacksonma

Chapter 5: Composition of Descriptions: A Progress Report
© AT&T Bell Laboratories 1993

Richard Kemmerer
Department of Computer Science, University of California at
Santa Barbara, Santa Barbara, CA 93106, USA
e-mail: kemm@cs.ucsb.edu

Chapter 10: Quality Assurance Working Group
© Richard Kemmerer 1993

Donald MacKenzie
Department of Sociology, University of Edinburgh,
18 Buccleuch Place, Edinburgh EH8 9LN, Scotland, UK
e-mail: D.MacKenzie@edinburgh.ac.uk

Chapter 3: The Social Negotiation of Proof: An Analysis and a
Further Prediction © Donald MacKenzie 1993

Peter Ryan
Chris Sennett
Defence Research Agency, St Andrew's Road, Malvern,
Worcestershire WR14 3PS, UK
e-mail: pyar@hermes.mod.uk/sennett@hermes.mod.uk

Preface, Chapter 11: Design Methods Working Group, Chapter
12: Conclusions, Appendix A: Survey of Formal Methods Tools,
Appendix B: Survey of Formal Methods Applications,
Appendix C: Acronyms and Trademarks © British Crown
copyright 1993

Margaret Tierney
Research Centre for Social Sciences, University of Edinburgh,
56 George Square, Edinburgh EH8 9JU, Scotland, UK
e-mail: M. Tierney@edinburgh.ac.uk

Chapter 2: Formal Methods of Software Development: Painted
into the Corner of High-Integrity Computing?
© Margaret Tierney 1993

John Wordsworth
IBM UK LABORATORIES Ltd, Hursley Park, Winchester,
Hampshire SO21 2JN, UK
e-mail: jbwords@winvmj.vnet.ibm.com

Chapter 7: Formal Methods and Product Documentation © IBM
Corporation 1993

Pamela Zave
AT & T Bell Laboratories, Room 2B-413, Murray Hill, NJ 07974, USA
e-mail: *pamela@research.att.com*

Chapter 5: Composition of Descriptions: A Progress Report